STEEL REMEMBERED

STEEL REMEMBERED

Photographs from the LTV Steel Collection

CHRISTOPHER J. DAWSON

THE KENT STATE UNIVERSITY PRESS Kent, Ohio

Published in Cooperation with The Western Reserve Historical Society

Library of Congress Catalog Card Number 2007021837
ISBN 978-0-87338-911-2
Manufactured in the United States of America

11 10 09 08 07 5 4 3 2 1

All photos are from the LTV Steel Collection, Western Reserve Historical Society, Cleveland, Ohio.

LIBRARY OF CONGRESS CATALOGING-IN-PUBLICATION DATA
Dawson, Christopher J., 1969–
Steel remembered : photographs from the LTV Steel collection / Christopher J. Dawson
p. cm.
Includes bibliographical references and index.
ISBN-13: 978-0-87338-911-2 (hardcover : alk. paper) ∞
1. Steel-works—Ohio—History—Sources. 2. Western Reserve Historical Society—Photograph collections.
3. LTV Steel Company—Archives. 4. Steel-works—Ohio—Pictorial works. I. Western Reserve Historical Society.
II. Title.
TN704.U52O354 2007
672.09771'022—dc22 2007021837

British Library Cataloging-in-Publication data are available.

CONTENTS

PREFACE

SAVING THE LTV COLLECTION

When historical societies acquire business records, it is often because executives from that business are on the board of the society, or the historical society has cultivated a relationship with the company, working to gain the company's trust and informing the company about the historic importance of its records. While that's a best-case scenario, it doesn't compare to taking in a collection over ham-and-cheese sandwiches and potato salad. That was the way the Western Reserve Historical Society (WRHS) acquired the LTV Steel Collection, not only one of the largest collections of steel industry records in America, but also the largest collection the Society has ever acquired. And it was all a matter of being in the right place at the right time.

Soon after LTV Steel was formed out of the merger of Jones and Laughlin (already owned by the LTV Corporation) and Cleveland-based Republic Steel, the new company declared bankruptcy, and the WRHS made an effort to save some early records, particularly from Republic Steel. The WRHS acquired only a few boxes of photographs and publications (many of which appear in this book), and further attempts to collect more records were not fruitful. As the years passed, the Society also tried to collect artifacts from the Cleveland mills operated by LTV, but, again, there was little interest on the corporation's part. Finally in the late 1990s, the WRHS embarked on a plan to design a new museum that would celebrate Cleveland's transportation and industrial history, which unfortunately did not pan out due to economic factors. When approached, LTV expressed interest in supporting the museum plans, especially as a sizeable exhibit on the history of steel in Cleveland was also planned. The first result of this new relationship was that the WRHS had access to LTV retirees for oral history interviews. Many of the retirees interviewed also worked as part-time tour guides of LTV's Cleveland Works, under the direction of Deb Kavulich, the LTV director of public affairs. Over time, WRHS staffers were permitted to assist the tour guides by providing historical content to the beginning and ending of each tour. At the time, I was the curator of Urban and Industrial History at WRHS, and one of my research interests was the steel industry, so I jumped at the chance to participate in the oral history recordings and help with the tours.

The retirees and Deb Kavulich saw WRHS's honest desire to preserve the history of the steel industry in Cleveland and worked to help us, because LTV Steel was again heading toward bankruptcy. One day in the summer of 2000, we were meeting with Deb when she casually asked if we'd be interested in some old photographs. Without hesitation, we said we were, and she took us into the basement of the LTV Cleveland Works headquarters, where she showed us a large file cabinet full of negatives, another cabinet filled with 35mm slides, and twenty-two boxes of photographs and negatives. All the images were either from LTV, Republic Steel, or Otis Steel, the pioneering Cleveland-based steel company that operated the west side mills of Cleveland Works, until it was purchased by Jones and Laughlin in 1942. We were shocked and amazed at this absolute treasure trove. Historians long to discover a find like this, and my breath was taken away by this rare find. Few if any photos existed of Otis Steel, one of the first mills in America to use the open hearth steelmaking process, and what we found from Republic were the negative files from its photographic department. There were tens of thousands of negatives, photos, and slides just sitting there. Deb knew that this material was at risk of being thrown away, so she made sure the company donated it to us. In fact, when she offered the material to the WRHS, we ended the meeting immediately so we could return that day with a van to pick up the boxes and file cabinets. We didn't want to risk missing out on such a treasure.

That initial discovery started the ball rolling. Deb began searching for other historical materials, and from her we received Otis Steel's ledger books up to 1942, when Jones and Laughlin purchased the company. She also found the ledger books kept by Corrigan, McKinney Steel, before it was purchased by Republic Steel. Deb found old issues of the Otis Steel employee newsletter and all sorts of other documents that historians consider priceless. As LTV slid deeper into bankruptcy, and it appeared that the company did not have a plan for emerging intact, much of the headquarters staff, including Deb, was laid off.

Deb, however, made sure that some of the remaining staff members knew what to do and she enlisted their help. First was Marty Kall, who also worked in the Cleveland Works headquarters. He found several boxes of old photographs and publications. The baton was then passed to Carol Schroeder, a longtime administrative assistant in the LTV corporate headquarters in downtown Cleveland. From Carol we got our biggest gift up to that point: twenty-seven oversized boxes of corporate publications, annual reports, photographs, and historic files. She also arranged for us to get industry-related bound periodicals from LTV's library and a number of boxes of corporate history files from LTV's legal and public relations departments. All told, we thought we were quite lucky and had gotten a small but excellent collection about LTV Steel and its predecessors. The jewel of the collection was the photographs, but the files added a great deal of context and would be a great boon to researchers. LTV appeared to be on its last legs, and we thought we had gotten what we could, especially once we learned that Carol Schroeder was next on the layoff list.

One day in spring 2002 Carol invited me out to LTV's technology center in Independence, Ohio. The center, once LTV's main research and metallurgical lab, was now the final headquarters of the dying company. I went out because the company was having a picnic for its remaining employees, most of whom would be laid off soon. Also planned was the opening of a time capsule, which had been laid in the cornerstone of the building when Republic Steel built it in the late 1950s. The building had been sold to a local developer, so company officials wanted to retrieve the time capsule. Alas, the time capsule became a fitting symbol of LTV; it was damaged beyond repair. Water had seeped into the protective case inside the cornerstone, and, though Republic Steel's engineers used high-quality stainless steel for the case, they had used a hemp gasket, which was not watertight. Water had filled the case, damaging all the contents. It appeared that the capsule had held all sorts of films, newspapers, and magazines, but all were beyond saving.

Despite this disappointment, the party ended up yielding a bigger time capsule, one that ultimately proved far more valuable. While we were eating our ham and cheese sandwiches, Carol introduced me to several gentlemen who were in charge of LTV's records management: Pat Wallace, Felix Albarano, and Phil Alverna. Their job was to oversee the destruction of LTV's records. The idea of destroying important historical documents was alarming to say the least, so I wasted no time in asking if the records could be donated to the WRHS rather than shredded. They agreed to talk it over with LTV's legal representatives. Almost immediately we heard back; LTV would be willing to turn over many of their records, which amounted to some forty-five thousand boxes.

That number was immense, and it represented probably about a fifth of the entire WRHS archives. It was impossible for us (or probably any archive in the country) to take the entire amount. However, the folks at LTV had an extensive box-inventory spreadsheet that they emailed to us. The spreadsheet itself was impressive; it came out to 325 pages, and the text was so tiny that we often had to view it with a magnifying glass. I pored over every line on the list, selecting what boxes I thought were important. Ben Blake, our assistant curator of manuscripts, who not only had a strong background in industrial and labor history, but was also a former steelworker, then looked over the list, adding what he thought was important for labor historians. Several other staff members made their suggestions as well. Ben and I had each done a good deal of research on steel-industry topics, so we both felt that we could pick materials relevant to the future needs of historians. Eventually we whittled the forty-five thousand boxes down to around four thousand. Thousands of boxes were actually filled with accounting spreadsheets that were utterly unreadable, because they were printed with proprietary software that had become obsolete. Those were easy to discard. The final box count was still a huge amount, but it was manageable. We even managed to take a few extra boxes that we neglected to tell our directors about, and Ben Blake did an admirable job of stashing them all over WRHS storage areas. We knew we had one chance to save as much as we could, so bending the rules seemed a prudent course. After processing, the collection would take up thirty-eight hundred lineal feet on the WRHS shelves. That is a big collection—more than half a mile long.

When going over the inventories, we did find some surprising things, one of which was finding fewer records from Cleveland-based Republic Steel than we had hoped for. The bulk of the non-LTV corporate records were records from Jones and Laughlin, nearly all of it relevant to Pittsburgh-area facilities. We did find a number of boxes relevant to Youngstown-area operations, especially Youngstown Sheet and Tube, but not as many from Cleveland as we would have liked. For a while we thought about only taking the Cleveland material. But the Pittsburgh material was so important to industrial and labor history that we felt we needed to save it as well, if not for us then for another archive. We had hoped to involve a Pittsburgh-area archive in our decision-making process, but we had limited time to make our choices. LTV was willing to deal with the WRHS only, and we had just a few weeks to save what we could before it went to the shredder. So we made the decision to save the important Pittsburgh materials, knowing that we could choose to keep them or donate them to a Pittsburgh-area archive. Boxes started arriving soon thereafter, and we began squeezing them in wherever we had room.

When we had a chance to finally see some of the things we were saving, we were astonished. There were original records and ledgers from Benjamin Jones, the founder of Jones and Laughlin; ledgers from his original partners, Bernard and John Lauth; letters to Benjamin Jones from Alexander Holley, the first real designer of steel mills; employee records, possibly even more than a hundred thousand individual records just from Jones and Laughlin; labor-relations records from a number of LTV's predecessors, including some rare files dealing with the unionization struggles of the 1930s at Jones and Laughlin, which few, if any, labor historians even knew existed; records from coal mines, lake shipping companies, and short-line railroads; pay ledgers from various mills; and records and ledgers dating back to the 1850s and the earliest iron mills west of the Alleghenies. In short we had a collection that covered not only the steel industry but also the industrialization and development of the United States from the early 1800s on. It was staggering, and as far as we could tell, it was one of the biggest collections of steel-industry material that had yet become available.

We knew that the collection was too large for us to keep and handle, so we came up with a plan to process the entire collection in order to make it available to researchers. This is crucial, because an archive needs to protect its collections as much as give access to them. A processed collection will have a catalog to allow researchers to find exactly what they need so they won't have to "paw" through and risk damaging documents. Besides, researchers would rather have a catalog than spend countless hours looking through unmarked and unorganized boxes.

We needed to come up with a processing plan, create a central catalog electronically and on paper, organize the collection geographically, and distribute certain components of the collection to other archives. Materials from the Youngstown area would go to the Ohio Historical Society's Center of Industry and Labor Youngstown Historical Center of Industry and Labor. Materials from the Pittsburgh area would go to the Historical Society of Western Pennsylvania's Heinz Pittsburgh Regional History Center, and so on. Eventually we were able to get ten other library/archives to partner with us in a consortium. Together we wrote a grant proposal to the National Endowment of the Humanities (NEH), and in spring 2004 NEH awarded a quarter-million-dollar grant to the consortium project, "Providing Access to Collections Documenting the History of the American Steel Industry."

Formulated and directed by the Society's Margaret Burzynski-Bays with help from Steve Doell, the Society's curator of manuscripts, the project seemed titanic in scope. There were so many boxes, files, drawings, employment cards, and records as yet undiscovered. But Margaret devised a processing plan that would be simple, efficient, speedy, and still protect the important papers in acid-free folders and boxes while organizing them for ease of research. Speed was of the essence; the NEH grant was for two years. If we couldn't process the collection in that time, we would have to seek additional funding, and that would be difficult (though we later found that a number of external donors were willing to step forward and help fund the project through to completion). A small staff of part-time and full-time temporary workers assisted Margaret. She taught them her method of processing the collection, and they quickly worked to complete their tasks. These processors included Hannah Kemp Severence, Timothy Beatty, Martha Boska, Monica Scullen, Scott Longert, and interns Alicia Miller and Joseph Wachtel. In addition, Benjamin Blake, who helped save the collection and who deserves a great deal of credit for securing the NEH grant, also helped with some processing. He and I had intended to help more with processing, but I ended up moving pallets of boxes with the Society forklift, creating more trouble than I was worth, and digging through piles of boxes to find documents and artifacts that could help us with our fund-raising efforts. Meanwhile Ben was swamped with his own work and could give the project only a little help.

Unfortunately, soon after the processing began, I was no longer an employee of the WRHS. Nonprofits were struggling in the weak economy, with grants and financial support eroding, so the Society had to cut a large amount of staff. Ben Blake and I were among those let go. Fortunately Margaret was able to stay (the NEH grant helped fund her position), and even in the face of various financial problems and staffing shortages, she and her crew persisted and ultimately accomplished the nearly impossible: getting the *entire* collection processed within the two years allotted by the grant, finishing in spring 2006. The collection then was divided among the WRHS, which retained all the Republic Steel and LTV Steel records; the Historical Society of Western Pennsylvania, which received all the Jones and Laughlin records; and the Ohio Historical Society, which received all the Youngstown Sheet and Tube records.

Future plans for the project include getting the catalog online so researchers can access this wonderful collection. The Republic Steel and LTV Steel records will be folded into the larger WRHS steel archives, which also include unprocessed collections of United Steel Workers locals and the LTV photograph collection that provided the photos for this book. Allie Miller, who was my summer intern before becoming a processor on the NEH project, sorted through the boxes and drawers of Republic Steel negatives and created a database, which revealed there were at least 43,512 negatives for Republic Steel alone. This did not include thousands of slides from Republic Steel and LTV or thousands of photos from Republic Steel, LTV, Otis Steel, or Corrigan, McKinney Steel. Processing and cataloging this collection will be another massive undertaking.

In addition to the efforts put forward by Margaret and her team, another major effort included raising funds to help process the collection. Led by the Society's former CEO and president Patrick Reymann, with crucial assistance from Tami Brown, Kelly Falcone, and Andy Junn, members of the WRHS Board of Trustees, and members of the WRHS Steel Advisory Committee, the campaign received funds from a variety of corporations, foundations, and individuals, including:

The Alfred P. Sloan Foundation
The Atlas Steel Company
The Helen B. and Charles M. White Charitable Trust
Cleveland Cliffs, Inc.
The George M. and Pamela S. Humphrey Fund
The Louise H. and David S. Ingalls Foundation
The Independent Steel Company
Mittal Steel
The Sherwick Fund of The Cleveland Foundation
The Singer Steel Corporation
The Society for Industrial Archaeology
United Steel Workers of America
The Livingston Mather Charitable Trust
The Elizabeth Ring Mather and William Gwinn Mather Fund
The Gordon and Llura Gund Foundation
Corning Chisholm
Dr. and Mrs. Kenneth R. Callahan
William E. Conway
Dr. John J. Grabowski
John Hollister
Mr. and Mrs. Elton Hoyt III
Peter and Donna Kelly
Kathrine and John Sargent
Thomas Tyrrell
Anonymous Contributors

These groups and individuals generously contributed funds so that the collection could be made accessible to researchers and the general public. Processing a manuscript collection is hardly glamorous work and often does not attract the number of donors that, say, a new building might. But in many cases, a manuscript collection can affect a greater audience and have a deeper impact than the finest new building. Ultimately many people came together to save an important collection of American history. It could have easily happened otherwise; the collection could have gone straight to the shredder, as is common with so many records nowadays. Most corporate legal offices recommend that records be destroyed to avoid the chance of litigation. But every document shredded is a piece of history. Surely some are worth saving.

We were extremely fortunate that Deb Kavulich, Marty Kall, Carol Schroeder, Pat Wallace, Felix Albarano, and Phil Alverna were willing to make the effort to help find historic records and artifacts for us, and help us acquire them from a dying company. That this book even exists is a testament to their dedication, appreciation for history, and pride in their industry's rich history.

. . .

The photographs in this book come from that initial LTV donation when the company entered its first bankruptcy in 1984 and from the photographs that we were shown in the LTV Cleveland Works basement in 2000. Those photographs would have probably been thrown away had Deb Kavulich not had the foresight to ask us if we were interested. Naturally we were, and you can see for yourself the end result, as this book displays some of the highlights from both wonderful photographic acquisitions. Unfortunately, as large as the combined photo collection is, there is still a lot missing. There are virtually no photographs from Youngstown Sheet and Tube and virtually none from Jones and Laughlin (though photos from both companies do exist in other archives). The bulk of the photos are from Republic Steel, but there are also thousands of photos from Otis Steel and Corrigan, McKinney Steel. Selections from those three companies make up the photos in this book.

And while there are indeed tens of thousands of negatives, prints, and slides in the collection, not all are of interest to the general public. Many images in the LTV Steel Collection are quite prosaic—a new piece of equipment being installed, an industrial process documented, an executive with a forced smile shaking the hand of a retiree he's never met before. But plenty of images capture much more, not only the incredible visual drama of the steelmaking process but also the human face of the industry, often together in wonderful compositions. For this book, images were selected that were not only good examples of industrial photography but also interesting to a general audience. They represent nearly all the phases of production, from the shipment of raw materials to the processing of finished products. The photographs are also unusual in that the few books on the market today with steel-industry images tend to focus on the decayed remains of closed steel mills or the mills being torn down; the photos in this collection show mills being built and the steel industry in the prime of its existence.

The majority of photographs in this book were taken from the 1930s to the 1960s either by Otis Steel– and Republic Steel–employed photographers or by Cleveland-area commercial photographers. Corrigan, McKinney photographs document the construction of that plant from 1913 to 1916. Where possible, the photographers have been identified, though in many cases that information is unknown. It is important to note that these were not well-known photographers. These were not the photographers hired by Roy Stryker to do the Farm Security Administration photographs during the Great Depression, many of whom went on to become famous. These were not well-known artists such as Margaret Bourke-White (who got her start photographing Otis Steel), prowling around the mills to create smoky and ethereal photographs. Nor were they photojournalists such as W. Eugene Smith, making an obsessive quest out of photographing Pittsburgh and its most famous industry. These were staff photographers, simply shooting photos for company publications. In many cases, they were not attempting to create "art," but in a select few they did, creating beautiful photos that compare with the finest photographs of the Farm Security Administration.

The following images capture a unique time and place in American history. These photos are quite literally a time capsule. Look closely at each one. While the eye often gets drawn to the brightness of the molten metal or the gigantic machinery, it is important to look beyond both and notice the details. Look at the people in the images, at once dwarfed by the process but also in control of it, though often just barely. Look at the faces, tired with the pressures of the job (when even a slip of attention can mean injury or death) but still resilient and defiant and even proud of their work. Look at the surroundings—building fixtures such as railings warped by the power of hot steel, debris and litter everywhere, messages chalked on walls or machines, old brick factory walls, and little to no safety gear. Remember that while few in these photographs wear safety helmets or goggles, this was not unusual in the 1930s. Helmets were generally not used until the 1950s, and even then some workers disliked safety gear because it wasn't "manly." Safety was not a big concern in the early decades of the industry, and worker mortality rates were high. Today, when safety concerns are an everyday part of our lives, we notice when the gear is absent. And these days, when "work" is primarily something done on a computer in a cubicle in an air-conditioned office building, the sort of work seen in these photos is often quite alien to the modern eye.

These photos capture an era that has vanished, and because the images are primarily meant for in-house company publications and annual reports, there is often little attempt to make the workers artificially heroic (as in advertisements) or to render the processes more mysterious or awesome than they already are. Yet there is still artistry in these images. The mill is a difficult place to take photos, even today. It is either too dark or too light, and often the light fluctuates so quickly and dramatically that it will give even the most sophisticated light meter fits. These photographers, especially the company-employed ones, most likely did not have the benefit of the extensive and intricate lighting setups or endless time to orchestrate the perfect shot. In the early photos from the 1930s, cameras and film were fairly slow, and often the people in the images are blurred because the light is too dim to allow action to be captured sharply. Despite these limitations the photographers created beautiful images of a bygone time and of an industry that today is but a shadow of its former self. These images bring us back to that era when America was at its industrial peak and steel was the undisputed king, a measuring stick of industrial prowess on an international scale. That time and those mills may be gone, but these photographs serve as a reminder that this era was not that long ago and is worth remembering and cherishing.

ACKNOWLEDGMENTS

Every book is a collaborative effort, and this book is no exception. My first level of thanks goes to the men and women at LTV Steel who worked to save this material, even as their own jobs were coming to an end. Without them, there wouldn't be an LTV Steel Collection or an NEH-funded processing project to make the material accessible. Thanks also to my former employer, the Western Reserve Historical Society, which was willing to take in such a wonderful collection and allowed me to proceed with this book. Dr. John Grabowski, the Society's director of research and Cleveland's leading historian, was absolutely instrumental in moving this project forward, even when my own inertia slowed or circumstances beyond my control threatened to sink the project altogether. He was my sounding board throughout the project and gave invaluable advice, as did my former supervisor, Dr. Edward Pershey, who not only supported the acquisition of the collection, but also insisted that I write this book to ensure these photos were shared with a wider audience. Credit must also be given to Patrick Reymann, former president and CEO of the WRHS, and Kermit Pike, chief operating officer. They allowed us to take in this collection and supported us as we pursued the NEH grant, which will allow the WRHS to eventually open the collection to the general public.

A very special thanks goes to another former WRHS staffer, Ben Blake, my partner in crime in acquiring the LTV Steel Collection. As my then-unpaid intern, Ben dropped everything to go to LTV and pick up the photo collection on the very day it was offered to us, which ultimately led us to acquiring this gigantic collection. Once he switched from unpaid intern to full-time worker in the WRHS library, Ben was one of the most important players in saving this collection. He pored over every inventory with me and, once the collection came in, deftly operated the Society's forklift, stashing thousands of boxes into the Society's already-jammed storage areas. If anyone is as fascinated by the steel industry and its rich heritage as I am, it's Ben. Without him, the saving and processing of the LTV archival collection wouldn't have gone forward the way it did.

Another person at WRHS who deserves countless pats on the back is Margaret Burzynski-Bays, who coordinated the processing of the LTV Steel Collection to perfection. Margaret was extremely patient when I raided the collection for materials for this book and when I hounded her endlessly for assistance, even though I was no longer employed at the WRHS. Steve Doell, the curator of manuscripts in the WRHS library, came into this project late but was completely sympathetic and supportive from the start. A much-needed thanks also goes to Kelly Falcone, who played a huge role in raising the money to process this collection, first with the NEH grant and then with external funding to match the NEH funds. From the start she kept telling me "you've got to write a book." I was listening, even when it seemed I wasn't. Once I finally began the book, Kelly tirelessly encouraged me and provided me with valuable information and feedback whenever I asked.

My next level of thanks is a personal one, to my parents, who have always supported me in any endeavor. Also thanks to my wife, Sarah, who generally doesn't understand my passion for steel mills but still lets me indulge it. And of course I owe a great deal to my grandfather, uncles, and various cousins who toiled in the mills to make a good living for our family and help propel us toward the American dream. When I was young, I didn't pay attention to the stories of their work in the mills. Now that I'm older, I'd love to be able to sit down, have a few beers with them, and talk about their lives and work, but unfortunately it's too late. I hope others will learn from my mistake and take the time to talk to their elders and learn the stories of their lives while they still can.

A final thanks goes to the men and women who made steel. This book is dedicated to them, because they worked tough, thankless jobs, helped make America what it is, and are too easily forgotten today. I hope this book shows how difficult, dirty, and dangerous the work in a steel mill truly was, and how the workers toiled in those mills to make the wondrous and absolutely crucial material that built this country.

CHAPTER ONE

MAKING STEEL

Steel is one of the most useful materials ever devised. It provides the infrastructure for our modern society: the skeletons of our buildings, the components of our automobiles, the bridges we drive across, and the rails for our trains. Eating utensils, computer components, needles and pins, paperclips, mailboxes, gardening tools, soup cans, guitar strings, and countless other items are all made of steel. Steel is used almost everywhere and for almost everything, even though it sometimes takes a backseat to plastic in consumer goods. Ever wonder what material the machines are made out of that extrude all those plastic water bottles? And in this era of "green" awareness, it should be noted that steel is completely recyclable. When a building is torn down or a car is sent to the junkyard all that steel goes back into the furnaces of a steel mill to be melted and born anew as fresh steel.

Steel doesn't occur naturally, however; it doesn't grow on trees and can't be dug up from the ground. What can be dug up is the raw material used to make steel. Although generally made in a mill, steel can also be made in complexes called "minimills" where scrap steel is melted into new steel. But for the sake of this book, the emphasis is on the old-fashioned steel mill, where the process begins with raw materials and ends with finished steel and steel products. The steel mill is a huge complex, encompassing hundreds of acres and often hundreds of buildings. Steelmaking requires an enormous amount of energy, water, and air and takes place in gigantic spaces filled with monstrous machinery. It is also a dangerous process. In recent years injuries and deaths have greatly decreased due to an emphasis on safety in mills.

This brief overview of the steelmaking process, which can be incredibly complex, will remain simple in nature, intended to provide background information to understand the photos in the book and the industrial processes they illustrate.

The process begins with raw materials. Steel is made primarily from iron ore, but coal and limestone also play a role. In the early days of the industry iron ore came primarily from Upper Michigan and Minnesota, where huge ranges of ore-laden properties were developed. By the 1950s those reserves were dwindling, and ore was brought in from Venezuela, Liberia, and Labrador, though ore continued to be mined in Upper Michigan and Minnesota and processed into taconite. The iron ore was mined, either underground or from giant pits on the surface, and then sent via ship to the steel mills, primarily in Indiana, Illinois, Ohio, New York, and Pennsylvania. Coal from Pennsylvania, Ohio, Illinois, and other states came in by rail. The coal was baked in airtight ovens to burn off impurities, leaving almost pure carbon. The carbon, or coke, was then combined with the iron ore and limestone in a giant bottle-shaped structure called a blast furnace.

In the next step air was heated in adjoining stoves and then released into the furnace. The raw materials were dumped in the top of the furnace, and as they worked their way down, the temperature increased due to the burning coke that melted the iron ore. The temperature inside the furnace could reach up to three thousand degrees, and the molten iron gathered at the bottom of the furnace. The calcium in the limestone attracted silicates and other impurities from the iron ore and formed "slag," a material that floated on top of the molten iron and was drained off periodically. Once blast furnaces were "blown in," or started, they ran continuously, sometimes for years, until it was time to shut them down to replace the heat-resistant materials lining the furnace interior. When the molten metal inside the furnace was deemed to be the right quality, the furnace was "tapped," allowing the molten iron to drain out. This happened four or five times a day. A plug of fire-resistant clay was knocked out of a hole near the bottom of the furnace, and the molten iron drained out of the furnace through channels carved in the dirt floor of the furnace casting house, a structure that surrounded the base of the blast furnace. Originally the molten iron flowed into the main channels and smaller channels off to the side; the iron cooled and solidified in those channels. In the early days iron workers felt that the smaller channels coming off the larger, main channel looked like piglets attached to a sow, and so the cooled iron product became known as "pig iron," a name still used today (and the individual pieces of cooled pig iron were often called "pigs").

In today's mills the molten iron generally goes through the main channel and into an insulated, bottle-shaped railcar, often called a "torpedo car." From there it goes off to the next process—steelmaking. Pig iron is too brittle for use, so it has to be refined further into either wrought iron (which is what a blacksmith uses) or steel. The big difference between iron and steel is the carbon content. Iron can be immensely strong—there are skyscrapers still standing with iron skeletons—but it can also be quite brittle. Steel is stronger than iron, but also flexible, so it is more suitable for a wide variety of uses.

One of the earliest methods of producing large amounts of steel commercially was the Bessemer process, developed by Sir Henry Bessemer in England during the 1850s. With the Bessemer converter, molten pig iron from the blast furnaces was poured into an egg-shaped vessel, which then tilted upward as a blast of air was blown through the hot metal. The oxygen provided heat to keep the metal in a molten state and served to blow out carbon and impurities. When the metal in the converter was judged to be the correct quality, the blow was turned off, the vessel was tilted back downward, and the steel was poured out, all within twenty-five minutes. This process was one of the most dramatic parts of the steelmaking process, as fire would leap out of the converter into the sky. At night, the glow from Bessemer converters in full blast could light up that area of the steel mill.

While it was dramatic and fast, the quality of the steel wasn't always the best.

Another steelmaking process, called open hearth, was developed roughly about the same time as the Bessemer process. The open hearth was a special, relatively shallow furnace that utilized the hot air and gasses for combustion. This method of steelmaking also used a good deal of scrap steel, which allowed for recycling of imperfect steel from the mill and scrap steel from any source, such as old railroad rails or automobiles. The metal scrap was put into boxes, which a large machine inserted and dumped into the furnace. Molten iron from the blast furnace was then poured in the open hearth furnace from a giant ladle held by an overhead crane, and limestone was added as for a flux, like in the blast furnace. The flames of the furnace played over the scrap and molten metal for around eleven hours, reducing the carbon content and impurities of the iron and creating steel. Because it was a slower process, open hearth allowed steel to easily be analyzed to determine its exact metallurgical properties, allowing for greater quality control. Once the steel was deemed "done," a tap hole at the rear of the furnace was breached, and the molten steel poured out of the furnace and into a waiting ladle.

Today neither process is used in the United States, though steel mills in Eastern Europe and South America still use Bessemer and open hearth. Most modern steel mills in America, Europe, and Japan use the basic oxygen process, which is similar to the old Bessemer process. A pear-shaped vessel is tilted sideways, and scrap iron and molten iron from the blast furnace are charged into it. Then the furnace is tilted upward and an oxygen lance is lowered in. Blowing in pure oxygen at supersonic speeds provides not only a fuel for combustion but also a method of stirring. Like the open hearth, samples can be taken repeatedly to ensure quality and exact metallurgical composition. Like the Bessemer, the process is also fast, with a full blow taking anywhere from thirty to fifty minutes.

A less common but still important process is the electric furnace. Initially used primarily for specialty steels, this process is also quite common in today's minimills, which create their steel from scrap, rather than from raw materials like traditional steel mills. This process is essentially a large-scale recycling effort. In an electric furnace, scrap is placed into a small furnace, which is then sealed; electrodes are lowered into the furnace, close to the surface of the metal, and an electrical current is sent through the electrodes and through the metal, melting it in a dramatic fashion. When the charge is complete, the furnace is tapped, and the steel flows into a ladle. In some cases the furnaces are tipped mechanically to drain the molten steel.

However the steel is made, the next step in most mills is to shape the steel. For many years in the twentieth century, molten steel was poured into ingot molds, forming ingots. This was done by taking the molten steel from the Bessemer or open hearth, pouring it into a ladle, and then slowly pouring out small amounts into waiting molds (a process called "teeming"), much like pouring water into an ice-cube tray. When the molten steel cooled, it solidified, and the molds were stripped off, leaving a solid steel ingot, which could be stored for future use or sold to a processor. In many cases the ingots were stored or sent on to the next process, which involved changing the shape of the ingot into something more useable. The ingots were lowered into a furnace called a soaking pit and heated to an appropriate temperature. The ingot was then sent to the blooming mill, where it passed back and forth through a set of rollers. On each pass the rollers were set closer together, squeezing the ingot further and reducing its thickness until it formed a "bloom," also called a billet or a slab. The blooms were stored, sent off to other processors, or further shaped in the mill by being reheated and sent through additional rollers. Depending on the rollers, they could force the metal into railroad rails, I-beams, or other structural pieces. Many of the blooms were sent through a series of rolling mills that would roll the steel into long, thin sheets. The initial bloom, which could start at six inches thick, could be rolled thinner than 0.05 inches and elongated from 13 feet to 1,210 feet. Once rolled that thin, the long sheets were then coiled up and sent to manufacturing plants, such as automotive plants. Sheet steel is extremely common today, used in automobiles, appliances, and countless other applications.

Steel mills today no longer pour ingots and roll blooms. The process has been greatly streamlined so that molten steel from the basic oxygen furnace is slowly poured into a continuous caster, where it passes through water-cooled rolls, which form the molten steel into thin slabs, ready for further processing. This innovation, which saves time, energy, and costs, is common in the major American steel mills today. In the heyday of American steel, during the early to middle of the twentieth century, the major steel mills rolled out a huge variety of shapes of steel and produced specialty bars, pipes, tinplate, special alloy steels, and forgings. However, today most of the major steel mills (other than the minimills) produce only a few basic products. Intermediary companies, called steel processors, now produce many of the specialty shapes or products once produced by the major steel companies. They buy the sheet steel or slabs and process them further for the customer.

One of the most visible changes in the steel industry today is the workforce. The amount of manpower necessary to make steel has greatly decreased, while technology and automation have allowed productivity to greatly increase. A good case in point is the Mittal Steel Mill in Cleveland's Cuyahoga Valley. Once two separate steel mills, Otis Steel and Corrigan, McKinney, and employing more than seventeen thousand workers, the mill today is a single entity and employs not even a tenth of that number, yet the smaller workforce produces more steel of a higher quality. This is not unusual, because the history of the steel industry includes technological advances, consolidation of production facilities, and reduction of workforce. While many in the industry look ahead, it's also worth looking back to see where the industry has been and the path it has taken.

In January 1937, the Cleveland Press *ran a series of articles about the steelmaking process and the mills along the Cuyahoga River. The language in these articles is wonderfully descriptive, and conveys portions of the steelmaking process in an especially dramatic fashion. This article follows the making of steel in the open hearth furnaces of Republic Steel's Corrigan, McKinney complex.*

"Fiery Sea Laps at Metal" *Cleveland Press,* January 4, 1937

Furnace No. 13, seething internally with 135 tons of molten metal, roared like an angry monster, licking out hot tongues of brilliant flame through the ports in each of its four doors.

Its withering heat rose in astigmatic waves to join the black smoke weaving amongst the girders overhead. Its basso clamor was like the booming of surf.

Half a dozen men moved about the doors, their attitude expectant and tense. They ran long steel rods through the ports and probed the fluid metal inside. They adjusted blue-stained glasses and peered intently at the bubbling mass.

Beneath No. 2 door a "boil" had formed—an adhesion of white-hot particles that were eating through the fire brick. There was fear that the fluid metal would come gushing out on the floor. A rare occurrence, yet one which must be guarded against.

Forty minutes remained until the "tapping of the heat"—removal of the "clay" (dolomite) plug from the rear of the furnace to permit the steel to hiss free. The ladle was ready and spouts had been "mudded up." But would the front of the furnace hold?

The workmen hovered, peered, conferred. A gigantic crane rumbled by overhead, a gigantic black spider crawling against the steel girders at the top of the plant. The radiance of the furnace ports increased, became blindingly intense.

For nearly 12 hours this 135 tons of metal had been boiling furiously in No. 13 open hearth furnace of the Corrigan-McKinney steel plant. Its temperature had risen to 2600 degrees Fahrenheit. It looked like white liquid flame.

Now, with the tapping time all but arrived, it had developed sudden willfullness and was threatening a leak. Not danger, but a costly loss of the heat was involved. Before this the metal has seared through a furnace front and there is nothing that can be done. The men retreat. The steel rolls out over the floor in a lurid sheet, melting everything it touches, even the embedded railroad tracks.

Furnace 13 is but one of 14 open hearth furnaces in the Corrigan-McKinney plant, which is a vast rectangular shed several hundred feet in length. Twice daily each furnace is tapped, and steelmaking offers little that is more spectacular than this periodic act.

The plant itself seems a covered amphitheater for the staging of some infernal show. Its high roof scales up to meet in a ridge above the bridge-work of beams. It is perpetually murky with smoke. Tracks sunk level with the earthen floor reflect the furnace glare, scintillating like silken ribbons stained red.

Phantom-like amidst fire and smoke, men and machines move through the plant. Furnace doors roll slowly upward; men are sharply silhouetted against the dazzling light. Cranes lumber along overhead embracing 50-ton ladles of liquid metal in their mechanical arms.

Well down the line from No. 13 is a furnace which was tapped a few moments ago. Through one open door it gives off a softer glow, the tempera scarlet of a storm sunset. It is waiting to be "charged" with the ingredients of the next heat, or batch.

Sidewise, resembling a mechanical crab, the "charging machine" trundles up. Its operator stands on a platform shielded from the furnace heat by steel and glass. He operates switches which guide the machine with much snapping and popping of electric sparks.

The round ram of the instrument picks up a shallow tray of scrap, thrusts it through the furnace door and inverts it, withdraws the tray and takes up the next.

Shortly the charging with scrap is done. A bell jangles somewhere up near the roof, warning workmen to one side. The crane approaches from the "mixer" with a ladle of molten iron. It is poured into the furnace like cake batter into a pan.

Furnace 13 went through this same charging almost a dozen hours ago, and the ingredients of steel had boiled in its interior ever since. Only 20 minutes now to the tapping, and it seems to quake with its own heat.

"Ever seen the inside?" asked my guide. "Come on and take a look."

He handed me his blue spectacles and led the way to one of the doors. I had expected it to be hot, but was unprepared for the real heat. I could see nothing against the fierce glare of fire; I could almost feel my skin crack. Then I adjusted the glasses and bent down to look.

Through the circular port I saw blazing fuel—coke gas mixed with tar or oil—thundering into the furnace from a large jet. It billowed along the roof in luminous, hard-driven clouds. Beneath it the steel bubbled as violently as syrup brought to a boil. Its surface danced like water whipped into frenzy by rain.

The strange thing about open hearth furnaces is the fact that they heat down instead of up. Resembling huge bakers' ovens 15 feet wide and 36 feet long, they boil the 135 tons or more of steel in a pool about three feet deep. The fire is overhead, radiating its volcanic temperatures downward into the pool.

That churning mass in Furnace 13 already had an identity of its own. It was heat No. 713495, destined within a few days to become fenders for new automobiles.

Ten minutes until tapping. We returned to the other side of the plant. A sort of nervous quickness disclosed itself in the actions of the men. One poked what appeared to be an elongated dipper into the furnace and brought it out filled. He poured the liquid steel into a mold held in tongs by another man, and instantly it began throwing up a small fountain of sparks as bright as fireworks.

A moment later the molder knocked it out, already congealed into a glowing chunk. He cooled the chunk in a tank of water and with a sledgehammer broke it in half. Two others joined him, and all three studied the fracture, estimating the amount of carbon in the steel by its structure and grain. Steel differs from iron chiefly in the lesser amount of carbon it contains.

Then suddenly a workman seized a wooden mallet and hammered on a gong at one side. The brazen clanging rang down the length of the plant like a fire alarm. No. 13 was ready to be tapped. The front of the furnace had held.

It takes an agile man to pry out the "clay" plug from the furnace vent and start the metal's flow. Sometimes it comes with a roar, spurting a hundred feet through the air to shatter in white-hot drops against the opposite wall. Sometimes it overflows into the pit beneath the ladle and, if water has condensed in there, an explosion results. The workmen know better than to get too near.

"We'll stand back here," said my guide, and pointed to a sort of balcony at one side.

We watched the workman begin probing at the plug with a long rod, the end of which was red hot. We saw sparks flying, heard the sputter of air turned on to help dislodge the plug. Slowly at first, then quickening to a steady flow, the steel began to run.

Crimson lightning flashed through the shed, and remained. The faces of the workmen turned to mobile bronze. The metal coursed through its trough, crackling with venom and spitting off sparks. It plunged into the ladle and heaved and churned. Smoke fought upward in incandescent brick-red clouds.

There is one more step after this before the steel loses its liquid state. The ladle is lifted by a crane, and through a vent in its bottom the metal runs into the cannon-like ingot molds, there to solidify. But it is conquered metal then, doomed to subservient roles. Its fury and real magnificence are gone.

CHAPTER TWO

CORRIGAN, MCKINNEY STEEL

The McKinney Steel Co. long has borne a charmed name in the iron and steel trade. Its impregnably self-contained position, its complete independence of action, its widely scattered and valuable properties, and its reputed earning abilities are traditional.

—*Iron Trade Review,* May 28, 1925

When on September 25 last, Republic took over the $65,000,000 assets of Corrigan, McKinney Steel Co., "Finis" was written to a corporate history as tragic and sensational as any in the annals of American business.

—*Fortune,* December 1935

The Corrigan, McKinney Steel Company had its beginnings as a shipping company run by James Corrigan, often referred to as "Captain Jim." Corrigan, described in *Fortune* magazine as "a hard-driving, broad-shouldered, loud-swearing Irishman," got his start in the oil business. He was a friend of John D. Rockefeller's brother Frank and sold several refineries to the oil tycoon. With money from Standard Oil stock given to him by John D. Rockefeller, Corrigan invested in an iron mining company in Wisconsin. Corrigan also built up a fleet of ore-carrying ships, bought more mines, and shipped his ore to Cleveland and other lake ports, where it was sold to various industrial concerns. His company was initially called Dallibra, Corrigan & Company, and then Corrigan, Ives & Company. As such, it grew and prospered until the Panic of 1893, when it was put into receivership. Stories vary about what happened next, but the accepted story is that Judge Stevenson Burke of Cleveland, who handled the receivership case and became a partner in the firm, named Price McKinney as receiver. Soon after, the company became known as Corrigan, McKinney & Company. The "& Company" represented Judge Burke. McKinney's choice was not a coincidence. Coming from Toledo, where he had worked for a furnace manufacturer, McKinney also had worked in the purchasing department of the Hocking Valley Railroad, where he got to know Judge Burke. Burke liked McKinney so much that he sent him on a mission to Mexico to straighten out the management of some silver and gold mines that he had financial interests in.

Whatever his association with Burke, McKinney was a skilled and efficient manager. He put the company back into order. The company then prospered once again, becoming one of the largest independent mine owners and lake shippers in the business. In 1908 Jim Corrigan passed away from peritonitis, leaving control to McKinney. Corrigan's son, James W. Corrigan, received 40 percent of the company's stock (McKinney owned 30 percent and Burke owned 13.5 percent). Because of his son's hard-drinking, playboy lifestyle, the elder Corrigan had directed that his son's shares be held in trust until "Young Jim" turned forty, with McKinney acting as trustee.

McKinney expanded the company further and built or bought blast furnaces in Cleveland, New York, and Pennsylvania. Now the company could increase its market because it could sell the pig iron. Under his direction, the company also shifted into steel production. Selling pig iron proved difficult, however, as many of the major steelmakers were now operating their own furnaces. Between 1913 and 1916 the company erected a steel-producing plant in Ohio's Cuyahoga Valley. The new complex had blast furnaces, open hearth furnaces, a coke plant, and rolling mills. Wartime production of the company greatly added to its prosperity. In 1917 McKinney changed the company's name to McKinney Steel Company, which greatly infuriated young James Corrigan. Corrigan challenged McKinney for control of his stock and won in the courts. However, even though Corrigan held more stock than McKinney, McKinney had the support of Judge Burke. As a result Price McKinney remained in control of the company's day-to-day operations, while James Corrigan journeyed to London where he become a socialite, enjoying the nightlife of London and entertaining royalty with his wife, Laura.

By 1925 the Corrigan, McKinney Steel Company was one of the largest and most integrated of the independent steel companies in the United States. Integration was a key concept in the steel industry; to control costs it was believed that a steel company should not just produce steel but also own its own sources of raw materials and shipping, and even own processing companies that would buy steel from the parent corporation. Andrew Carnegie was one of the early masters of integration in steel, while John D. Rockefeller mastered integration in the oil business. Corrigan, McKinney Steel did pretty well in the integration game. The company's main facility was in Cleveland, where it had fourteen open hearth furnaces, four blast furnaces, numerous coke ovens, a rolling mill, and plenty of room for expansion. The company also owned two blast furnaces in Josephine, Pennsylvania, and one blast furnace in Charlotte, New York, and it leased a blast furnace in Scottdale, Pennsylvania. McKinney Steel had extensive iron ore holdings in the Lake Superior district, with six mines in Upper Michigan and five mines in Minnesota. Coal came from the Greasy Creek and Wolfpit mines, both located in Kentucky, and limestone came from Williamsburg, Pennsylvania. The only thing the company did not own was ore ships, which was ironic considering the company's origins in ore shipping.

McKinney Steel expanded in 1925, building new finishing mills, new 21-inch and 18-inch sheet bar mills, and new 10-inch and 12-inch merchant mills. Like Otis Steel, a competing steel mill across the Cuyahoga River, Corrigan, McKinney attempted to create a "family feeling" among its workers. Like Otis and other Cleveland-area companies, Corrigan, McKinney's workers participated in company-sponsored sports leagues. The company also owned a number of houses near its complex, which it leased to its employees, and, during one particularly good year, Corrigan, McKinney gave cash bonuses at Christmas to all employees. In some cases, the bonus was the equivalent of an employee's yearly salary. And even though the next year wasn't as successful, Price McKinney still doled out the same generous bonuses. When *Iron Trade* magazine wrote an article about the company in 1925, it noted that the company's history had "a record of almost fabulous success that spans from the morass of receivership to the dazzling heights represented by many millions of tangible assets" and correctly gave the credit for all this success to Price McKinney.

The year 1925 also saw a change in company leadership. James W. Corrigan returned from London and announced he had acquired a controlling interest in the company. (He had acquired another 13 percent of the company's stock from one of Judge Burke's heirs and now held a controlling interest in the company.) He took over the presidency of the company, and one of his first acts was to change the company's name back to Corrigan, McKinney Steel. McKinney "retired" and served briefly as a company director. But the struggle for the company had taken a toll on him, and on April 13, 1926, Price McKinney committed suicide. Corrigan was now on his own and struggled mightily to keep his company on the successful course that McKinney had charted for it. Corrigan had never had any interest in the company before, and his years of unceasing social engagements in London did not prepare him for leadership of a steel company. He did not remain in control for long. On January 23, 1928, he died of heart failure on the front steps of the Cleveland Athletic Company, leaving the steel company in the hands of four women: Mrs. Corrigan, Mrs. McKinney, Mrs. Burke, and the daughter of Judge Burke. Never before in the history of the steel industry was a large, integrated steel producer owned entirely by women. Their ownership would only last a few years, though, as the Cleveland-Cliffs shipping and mining company bought out Mrs. Corrigan. Handsomely paid, she returned to London and again began hosting lavish parties for top society and royalty. For a brief time in 1934 Cleveland-Cliffs was in talks with General Motors about the mill, and it was thought that GM would purchase the mill, use the steel it needed, and sell the surplus. But the growing Republic Steel Corporation, needing a mill on the Great Lakes and requiring greater steelmaking capacity, purchased Corrigan, McKinney Steel in 1935. Unfortunately for the McKinney family, another tragedy ensued. In 1935 young Price McKinney Jr., who had been working his way up in the company, also committed suicide. The tragic story of Jim Corrigan and Price McKinney even had a final cruel twist; enemies in life, they became neighbors in death—the two men are buried near each other in Cleveland's Lakeview Cemetery. The steel plant they fought over became a key part of Republic Steel and ultimately survived Republic Steel's merger with LTV Steel, two bankruptcies, the demise of LTV, restarting under ISG Steel, and finally a sale to Mittal Steel.

Today, more than ninety years after its construction, the steel plant built by Price McKinney still makes steel in Cleveland's Cuyahoga Valley.

. . .

The photos in this section come from a large photo album, "Construction Photographs of the Corrigan-McKinney Steel Plant, November 4, 1913–April 18, 1916," which was kept in Republic Steel's engineering department at the Cleveland District until it was donated to the WRHS after the 1984 merger with LTV. The 310 black-and-white 8x10 inch photographs were mounted on cloth, which was then bound into the album. Some years after the photos were taken, Republic employee Chuck Bayless of the Cleveland plant's engineering department wrote basic captions on the back of each album page, describing each picture. These captions provide the background information for the following photographs. Twenty-six photographs from 1914–1915 were selected for this section, along with an image of the mill taken in the 1930s, to show the "completed project." No photographer is credited for the album, though several photographs in the album are stamped with "Charles T. Meacham, Commercial Photographer, Cleveland." Since most photos were taken in the same position repeatedly over the two-and-a-half-year construction period and are all of similar quality, it is likely that Meacham took the majority of these images.

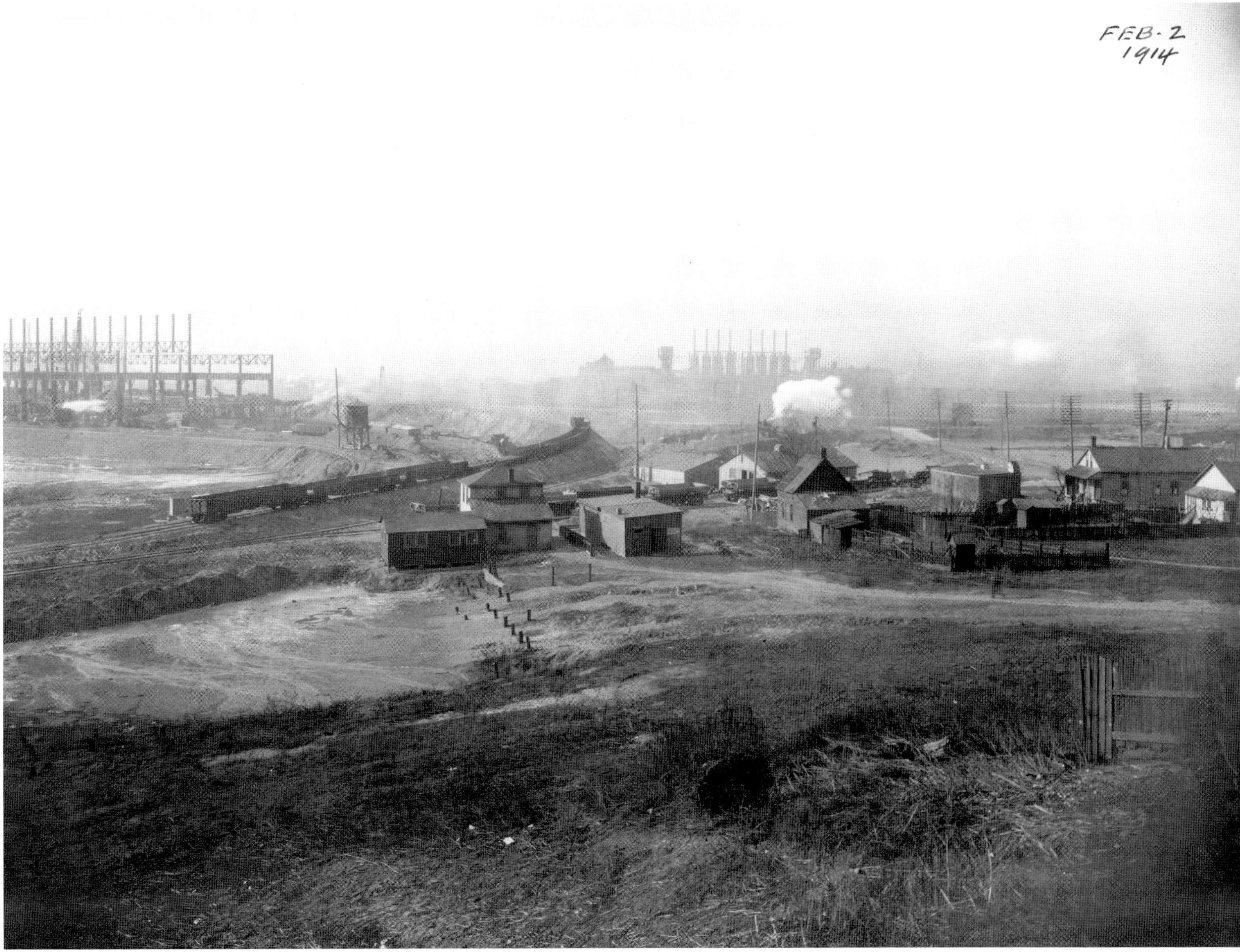

In this 1914 photo, Dille Road is in the right center; the photographer is looking southwest, with the open hearth building under construction to the left. In the center rear are Blast Furnaces 1 and 2 and ore docks.

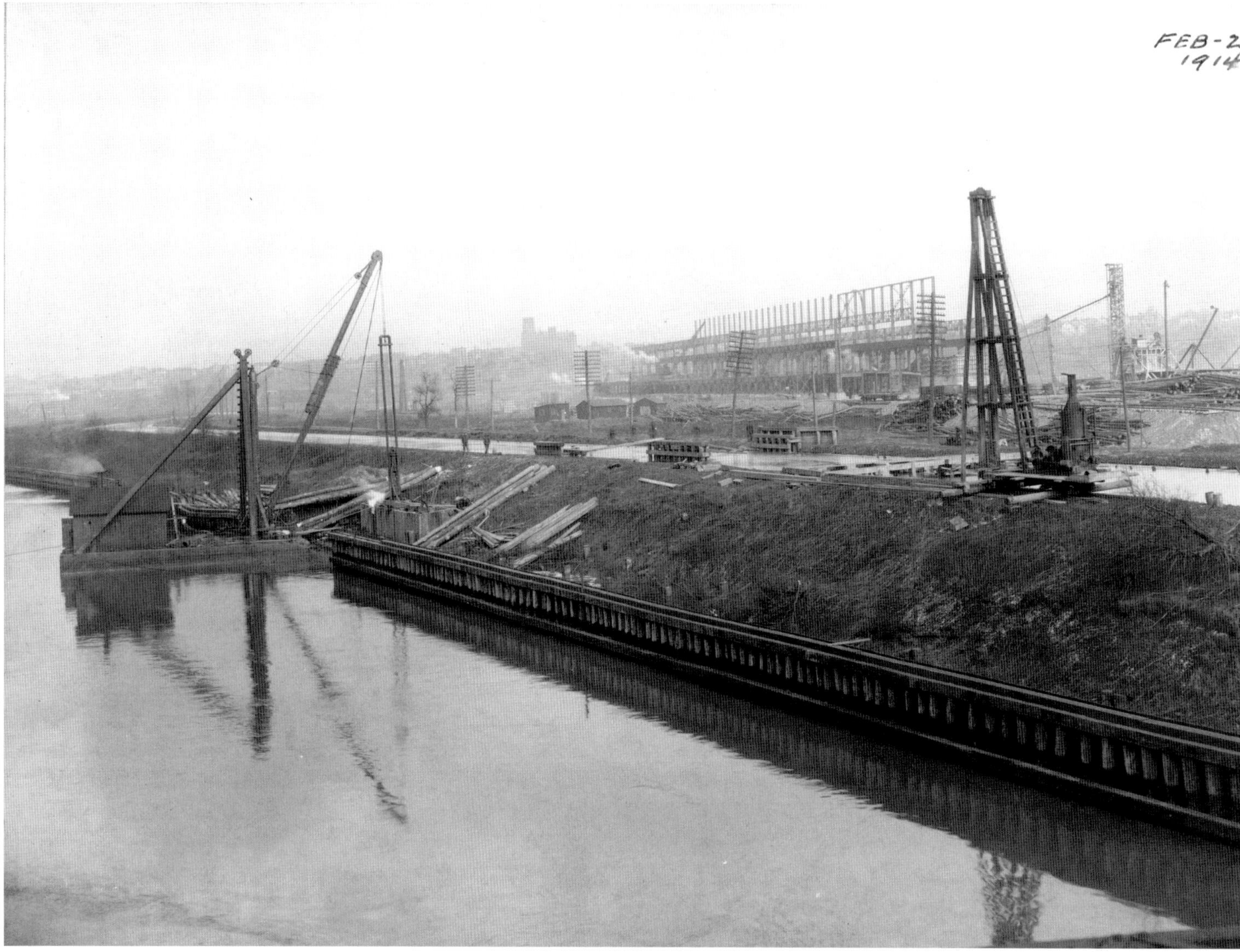

The photographer is looking northeast across the Cuyahoga River at the open hearth plant under construction. To the east of the river is the Ohio and Erie Canal, which essentially ceased operations in 1913, but was still filled with water one year later when this photo was taken. On the left, work begins on a bridge across the river for the River Terminal Railway, which was owned by Corrigan, McKinney. In the far distance is St. Wenceslas, a Catholic parish serving the Czech community. The church remained at East 37th and Broadway until it was demolished in 1963 for the construction of Interstate 77.

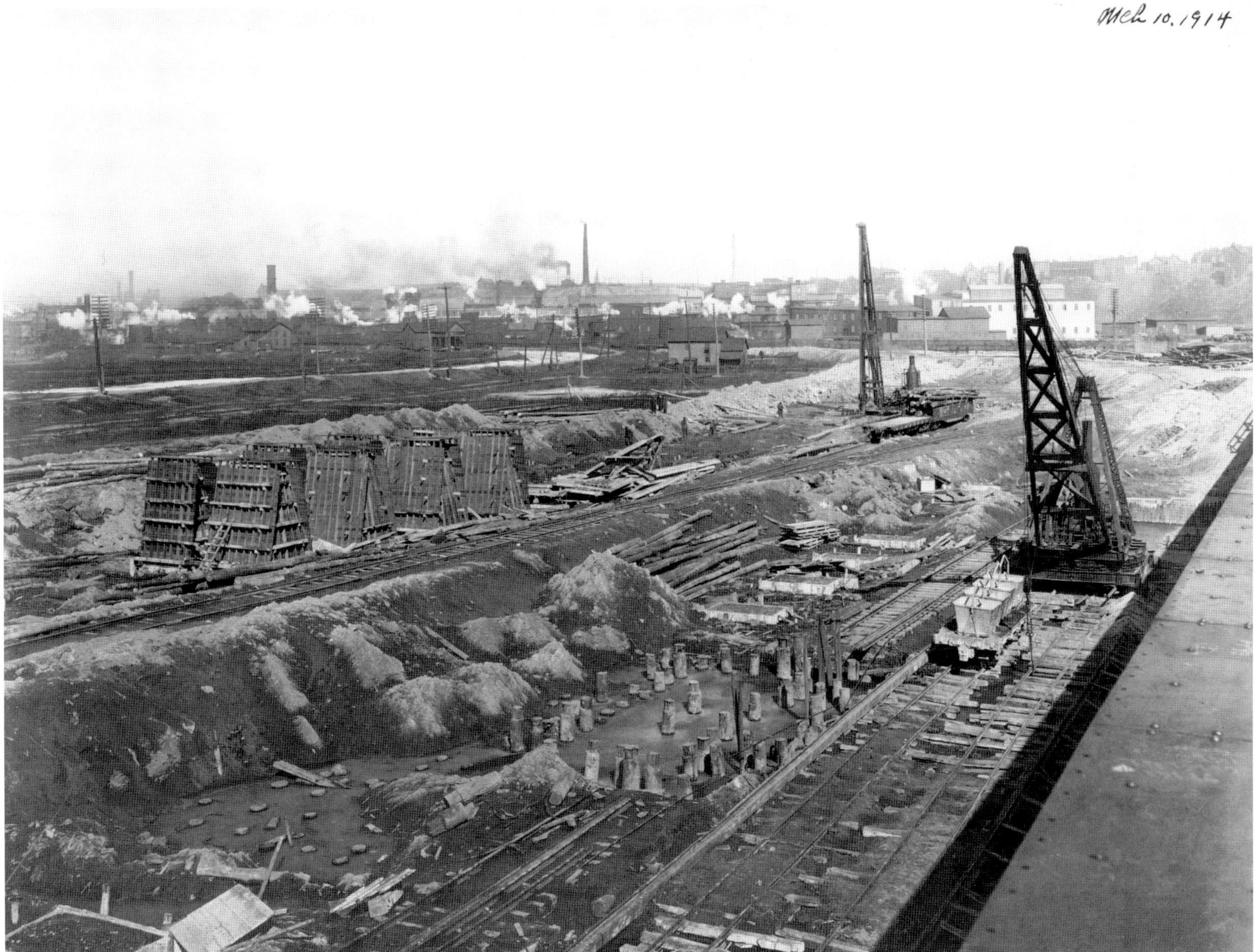

Looking north, with the Ohio and Erie Canal on the left and the stock yard foundations in the center. In the rear is smoke from other industries in Cleveland's Flats district.

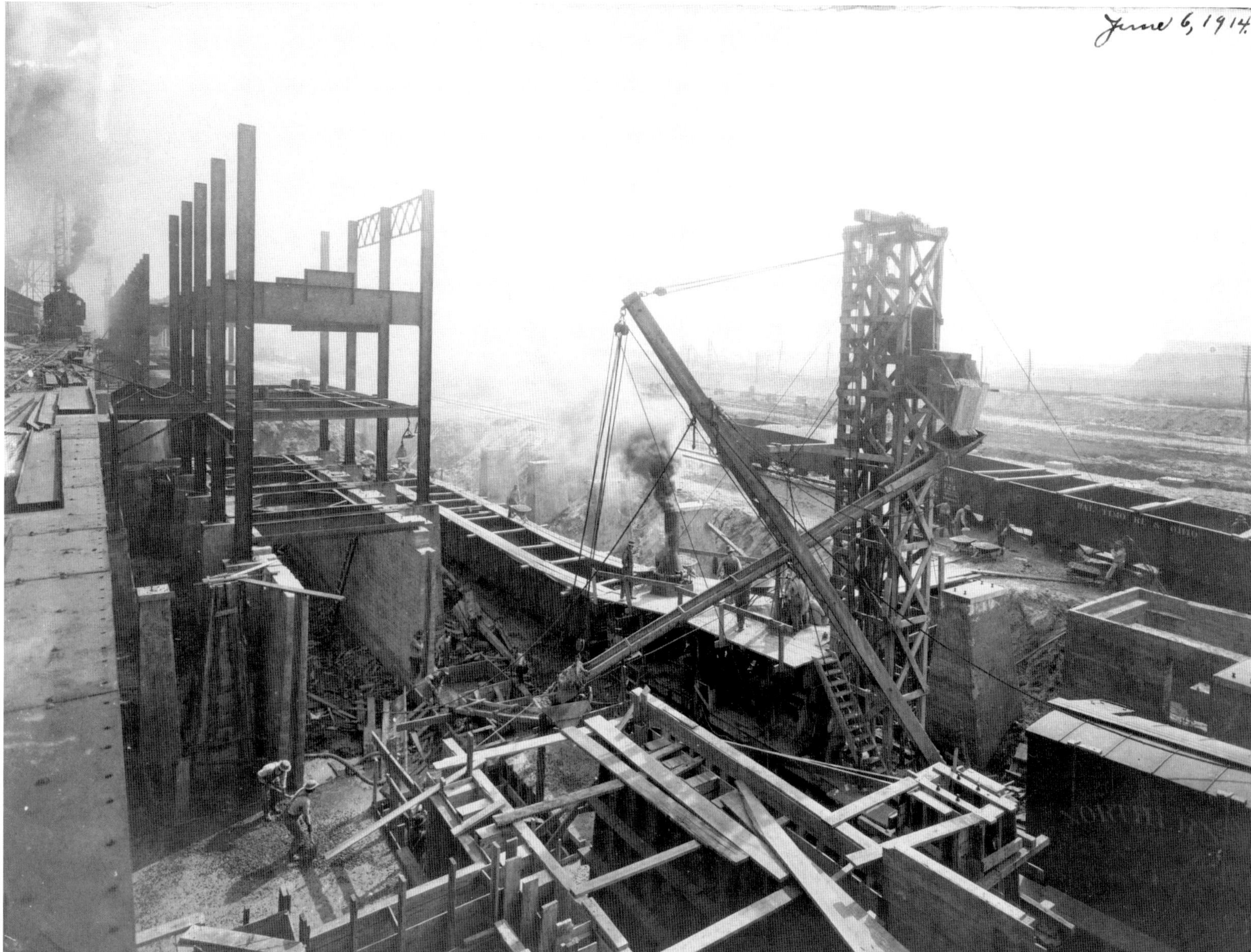

Looking southwest at the foundation for the plant's boilers. The wooden tower on the right is a cement mixer/pourer, moving around the work site on rails.

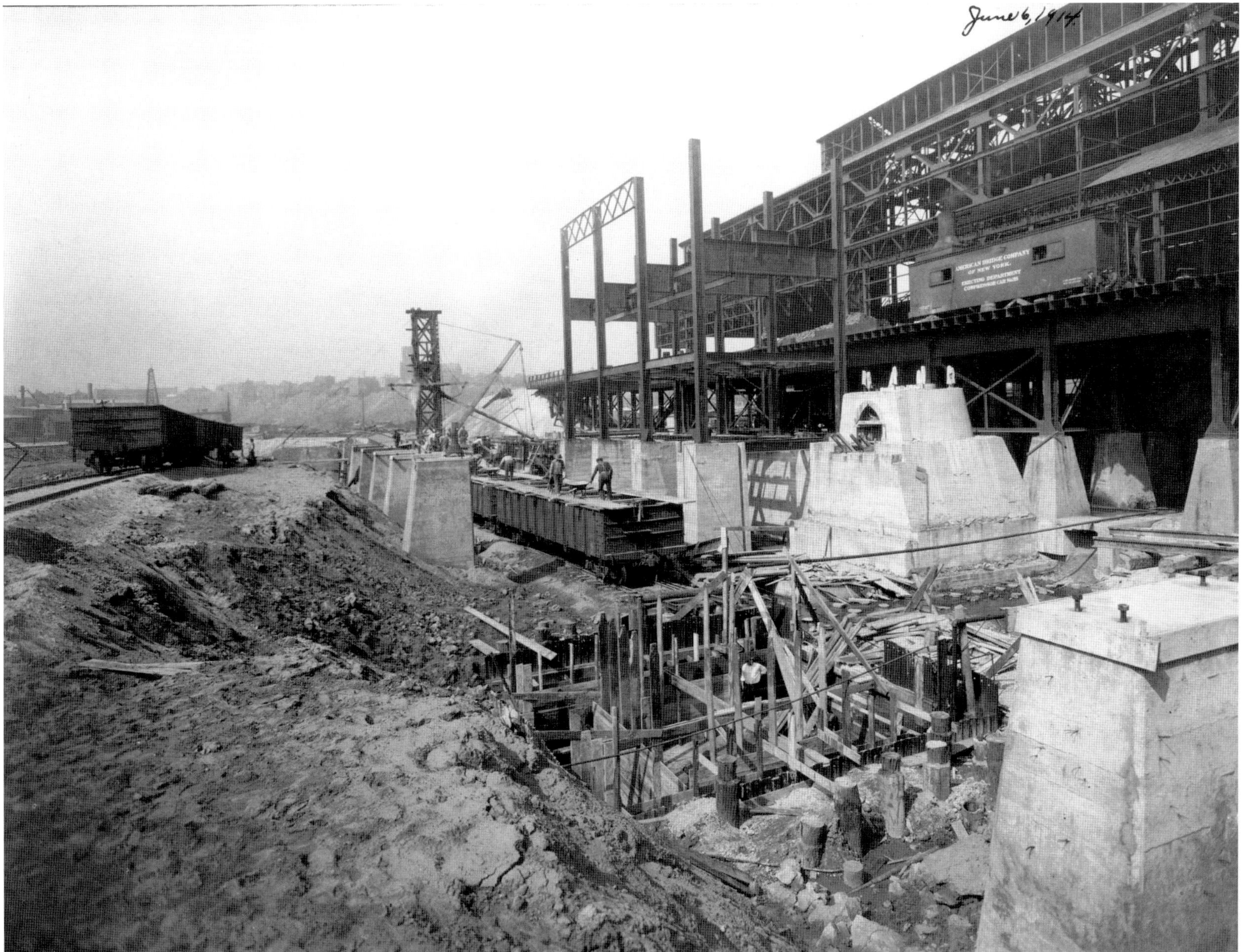

Looking north at the stock yard building foundation in 1914, with the open hearth building rising on the right.

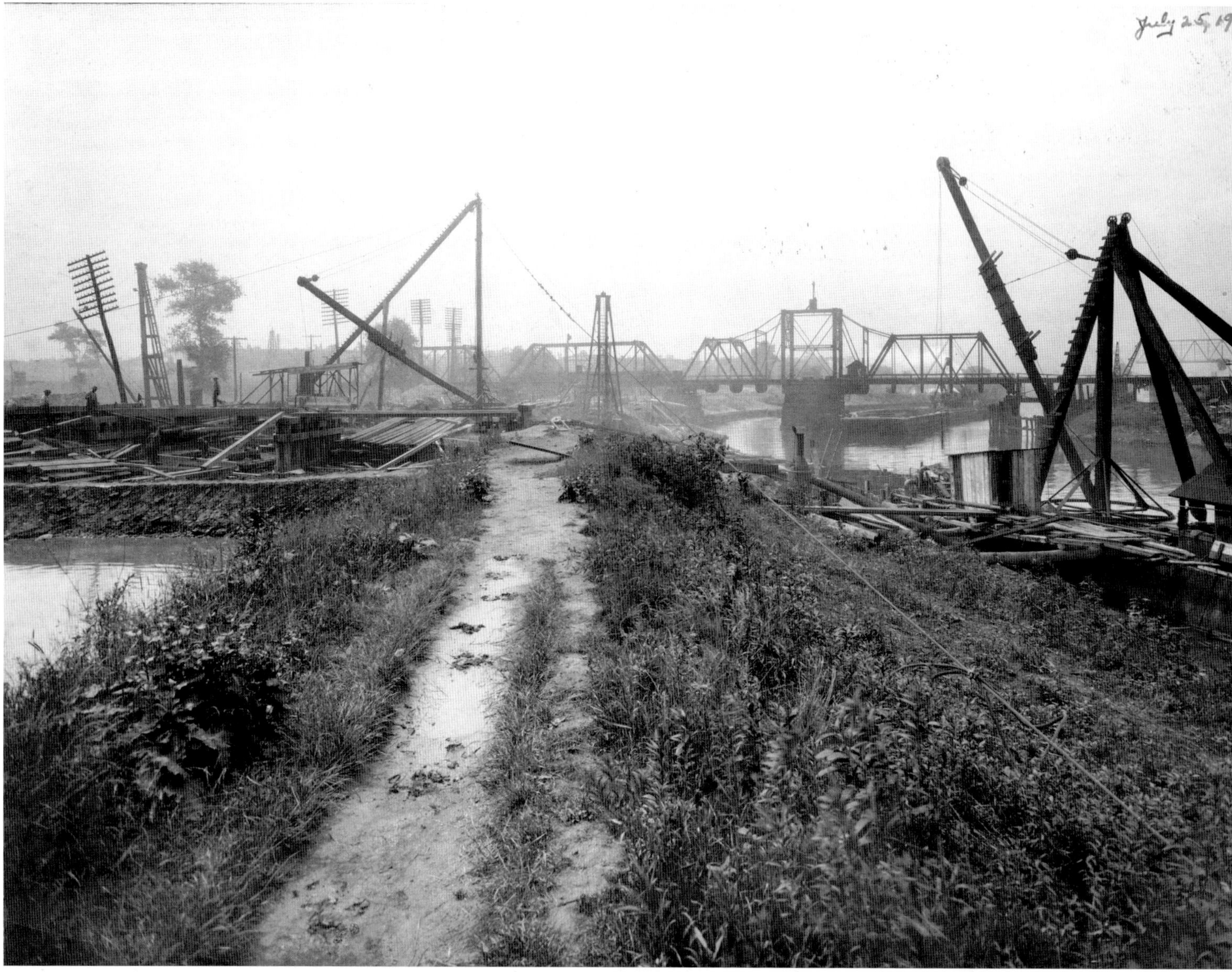

The photographer had his camera firmly planted on the Ohio and Erie Canal Tow Path for this 1914 photo, which is looking south. The canal is on the left, and the Cuyahoga River is on the right. Ahead is the site for the River Terminal Railway bridge to serve the open hearth. Note the dredge and barges coming up the river.

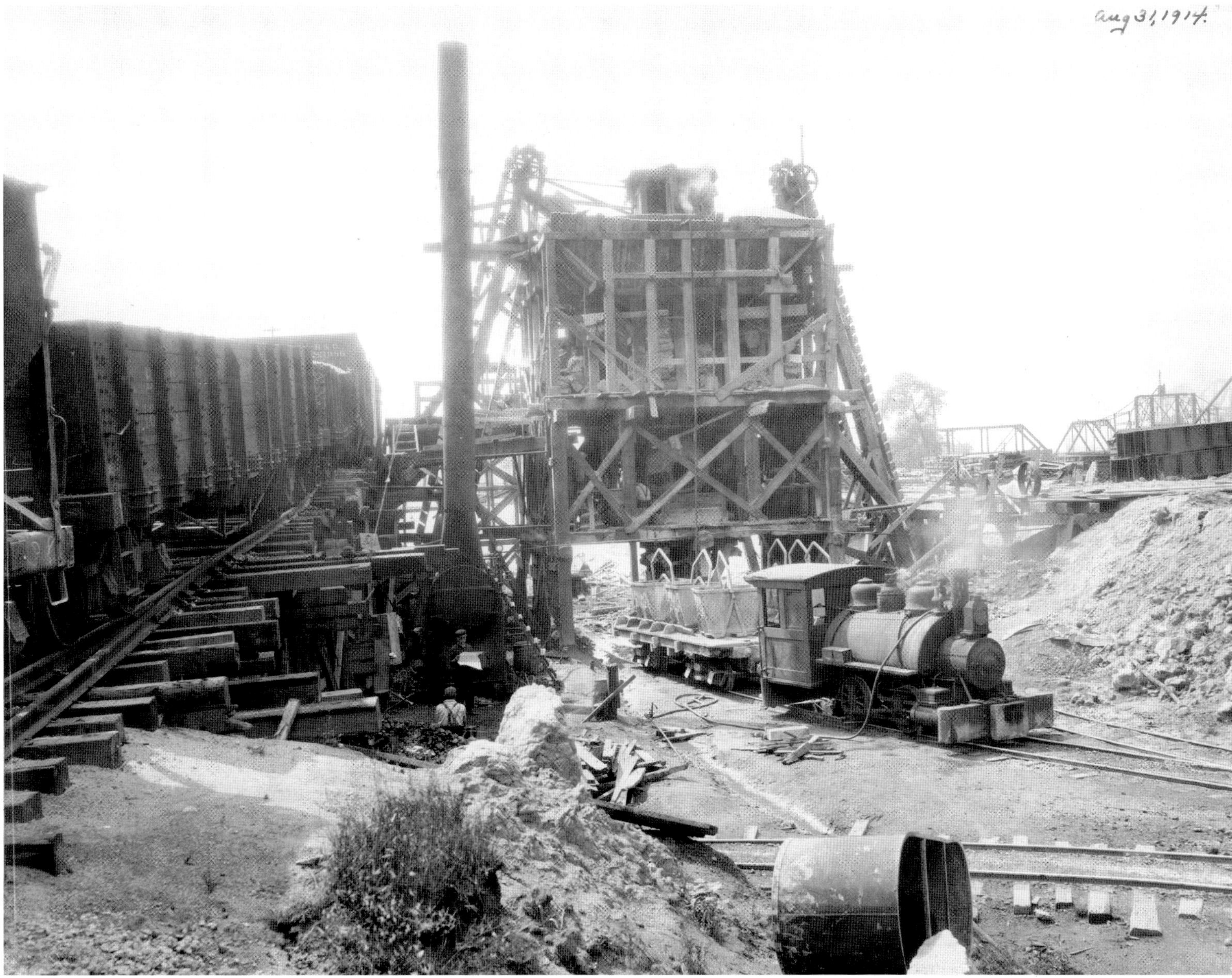

Looking south at the small 0-4-0 saddle tank switcher locomotive, which is under a large concrete mixer. To the left of the switcher, in the shadow of a steam boiler, is the engineer, reading the paper.

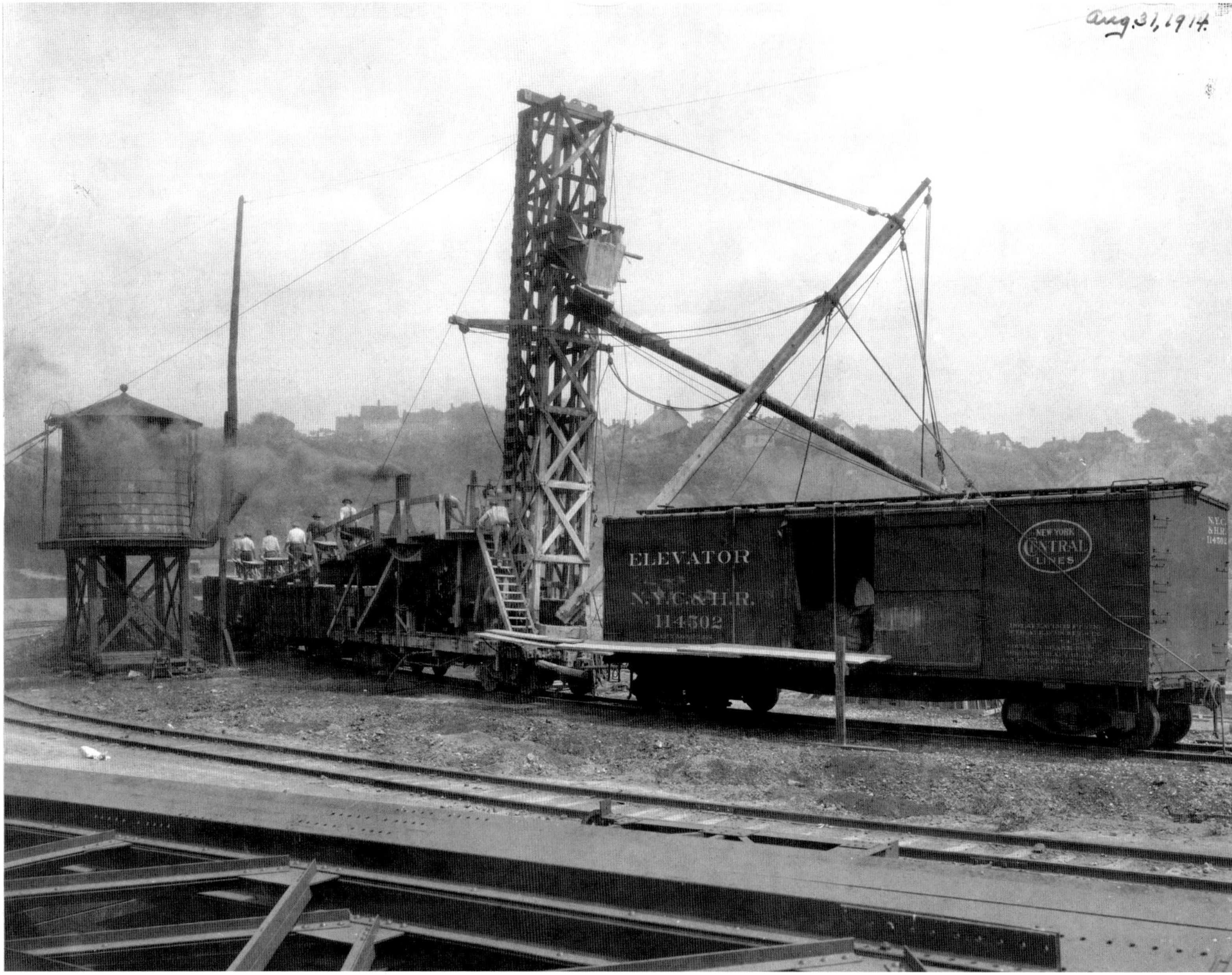

Looking northeast at the concrete mixer, which can be moved easily around the job site on rails.

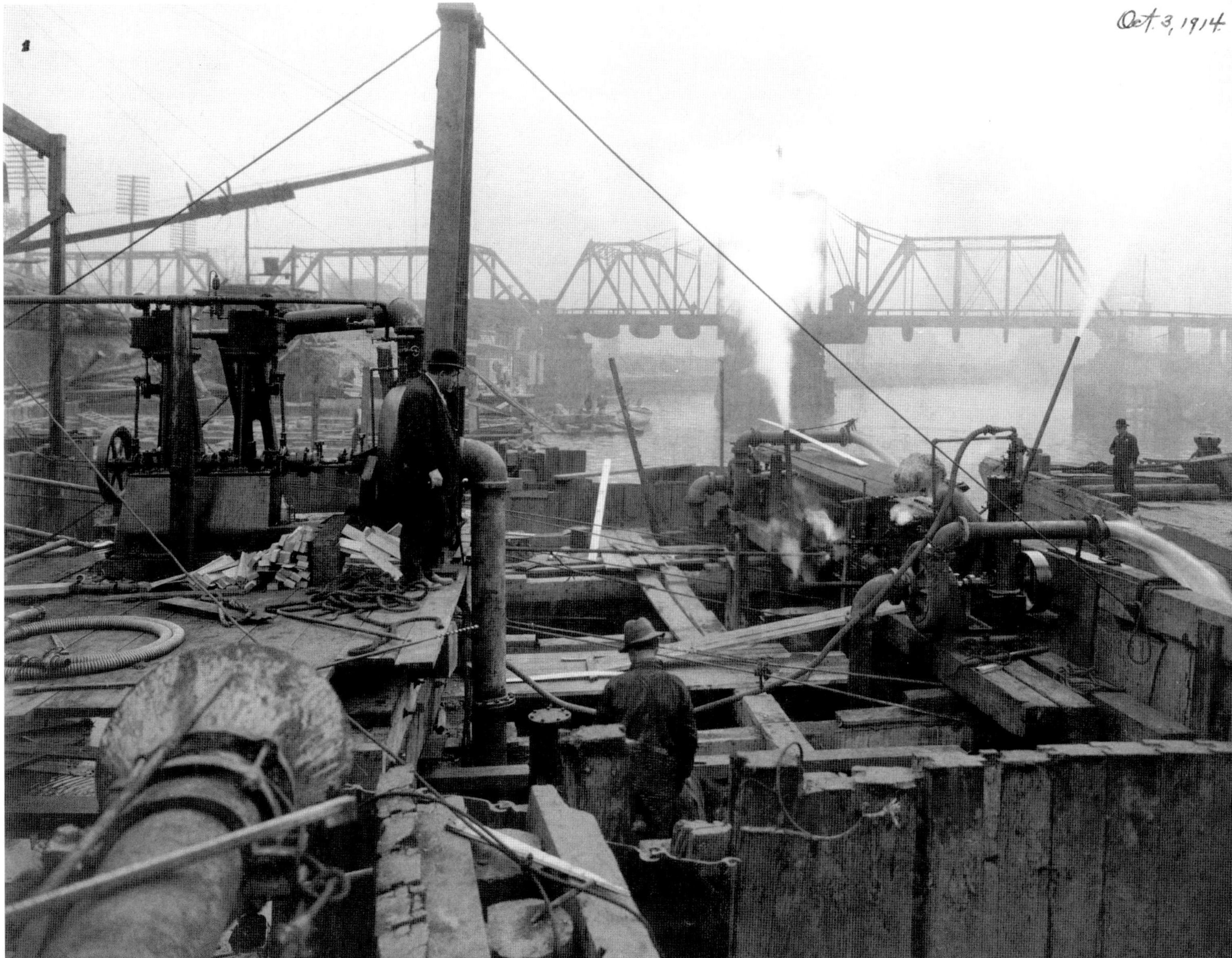

Looking southwest with the Cuyahoga River on the right. The work site is a cofferdam adjacent to the east bank of the river, site of the River Terminal Railway bridge.

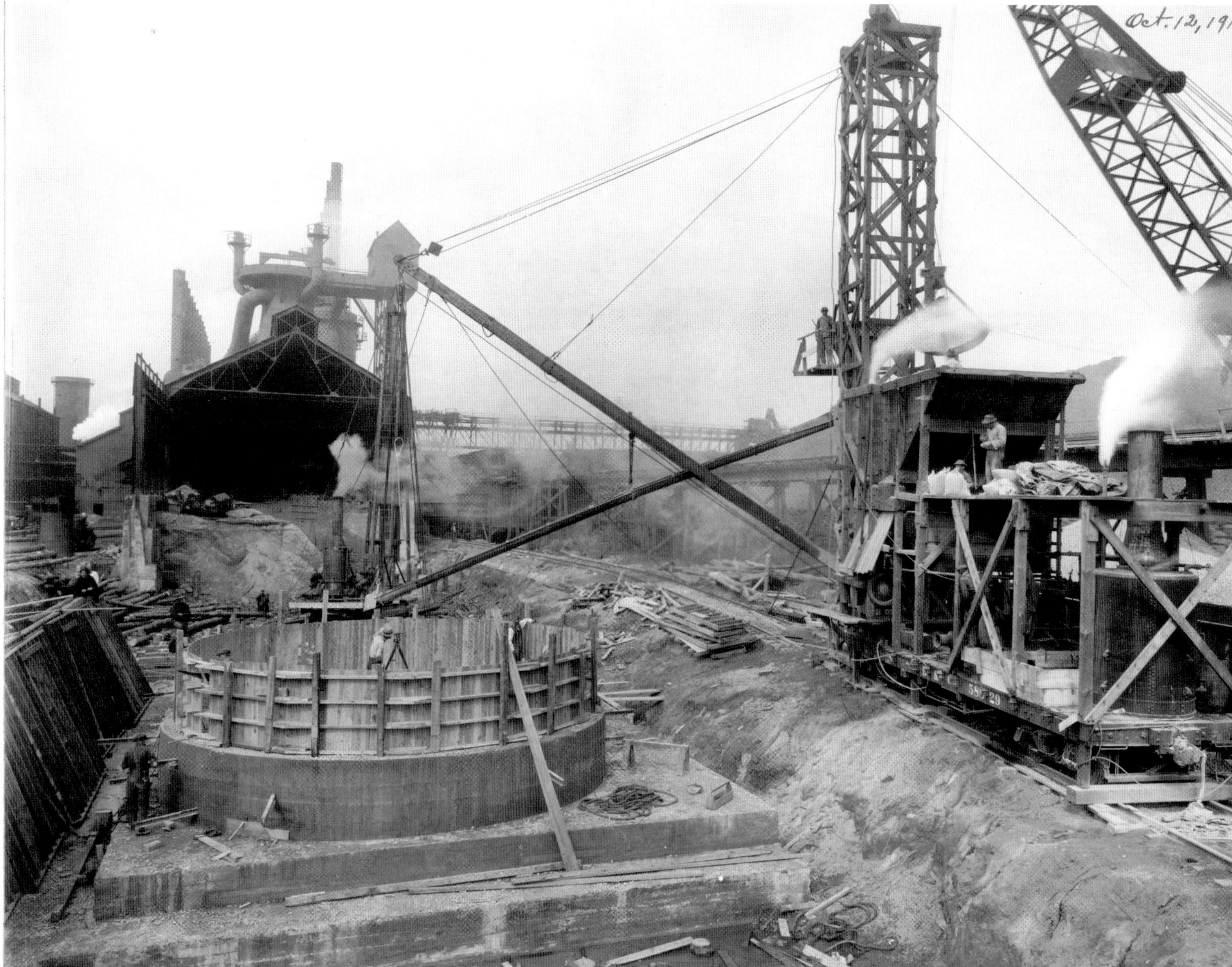

Looking northeast at Blast Furnaces 1 and 2 and at the site where Blast Furnace 3 is under construction.

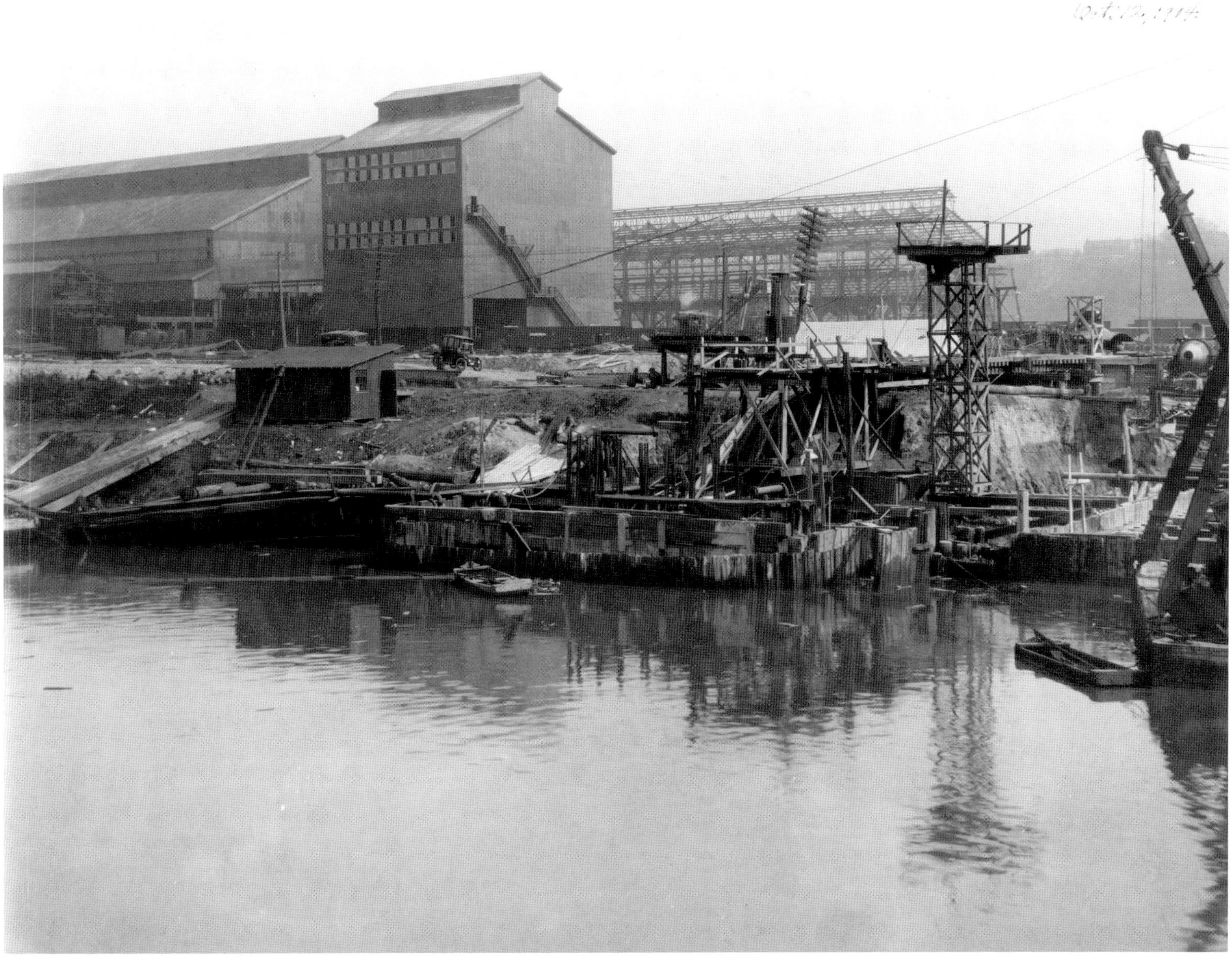

View of steel plant and east, from the west bank of the Cuyahoga River. On the right side construction for the eastern foundation of the railroad bridge continues. Slightly to the left of center is an electric automobile, most likely a Cleveland-built Baker Electric.

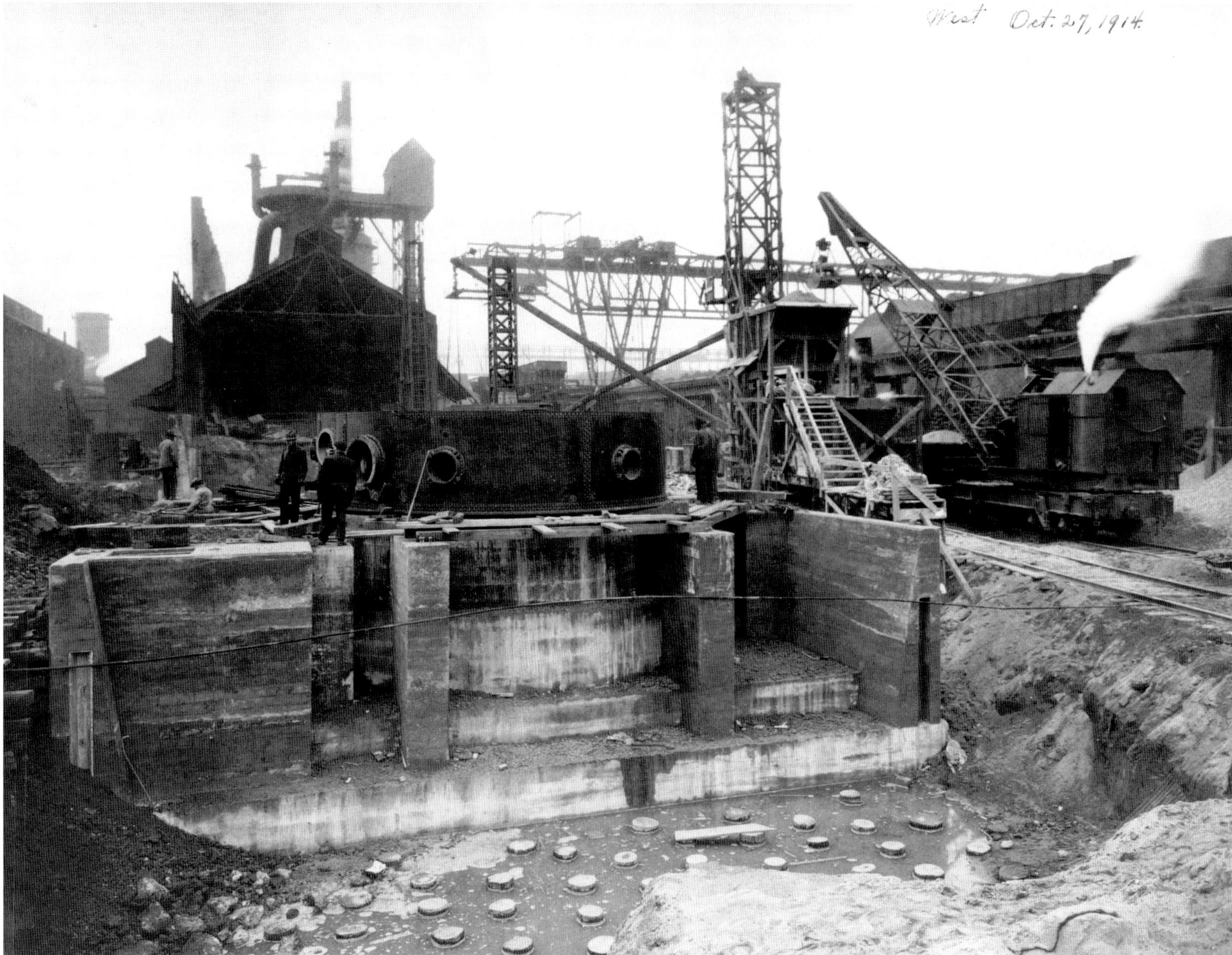

On the west side of the Cuyahoga River, looking north at the foundation of the hot air stoves for Blast Furnace 3.

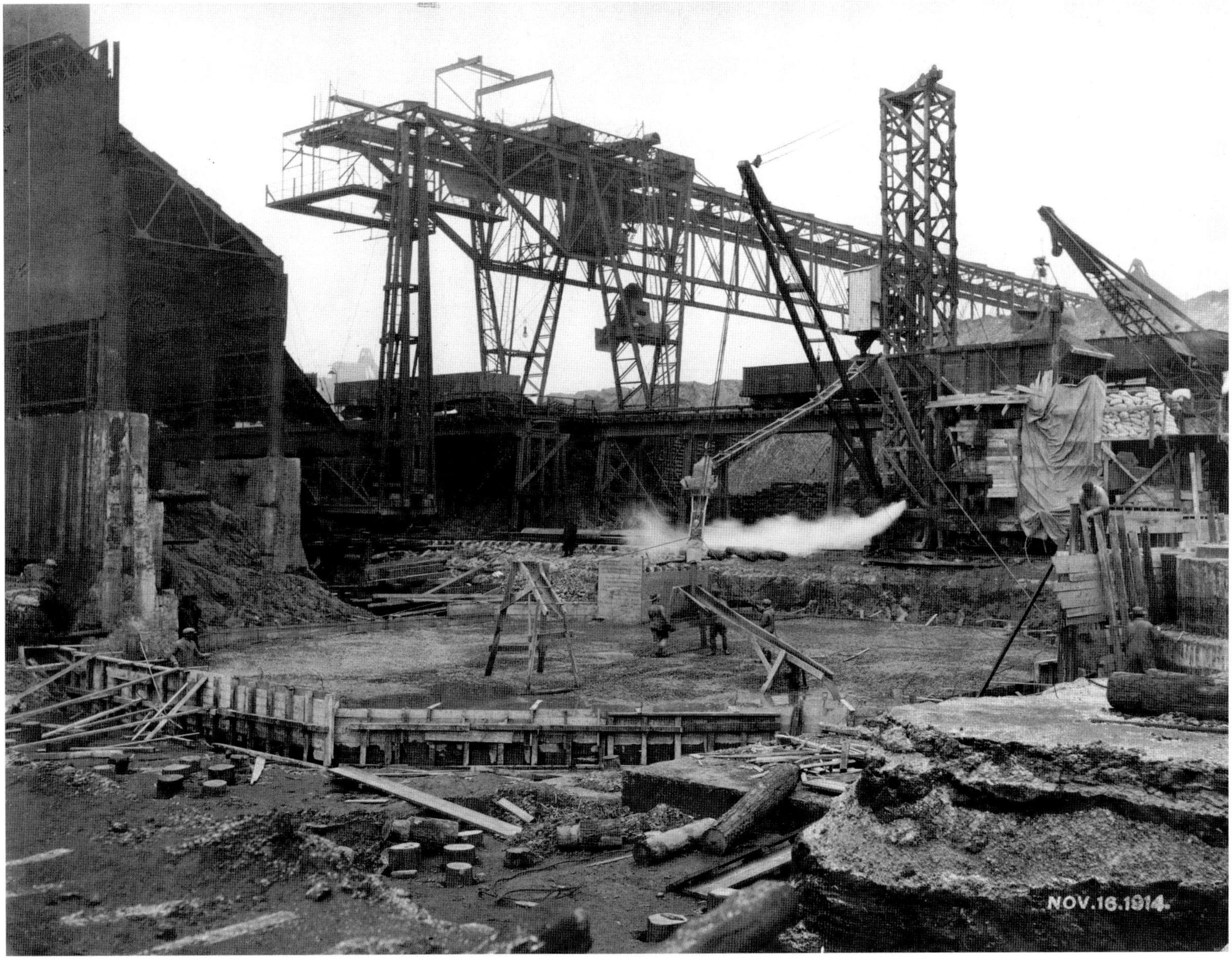

On the west side of the Cuyahoga River, looking northeast. In the foreground workers are pouring the foundation for Blast Furnace 3.

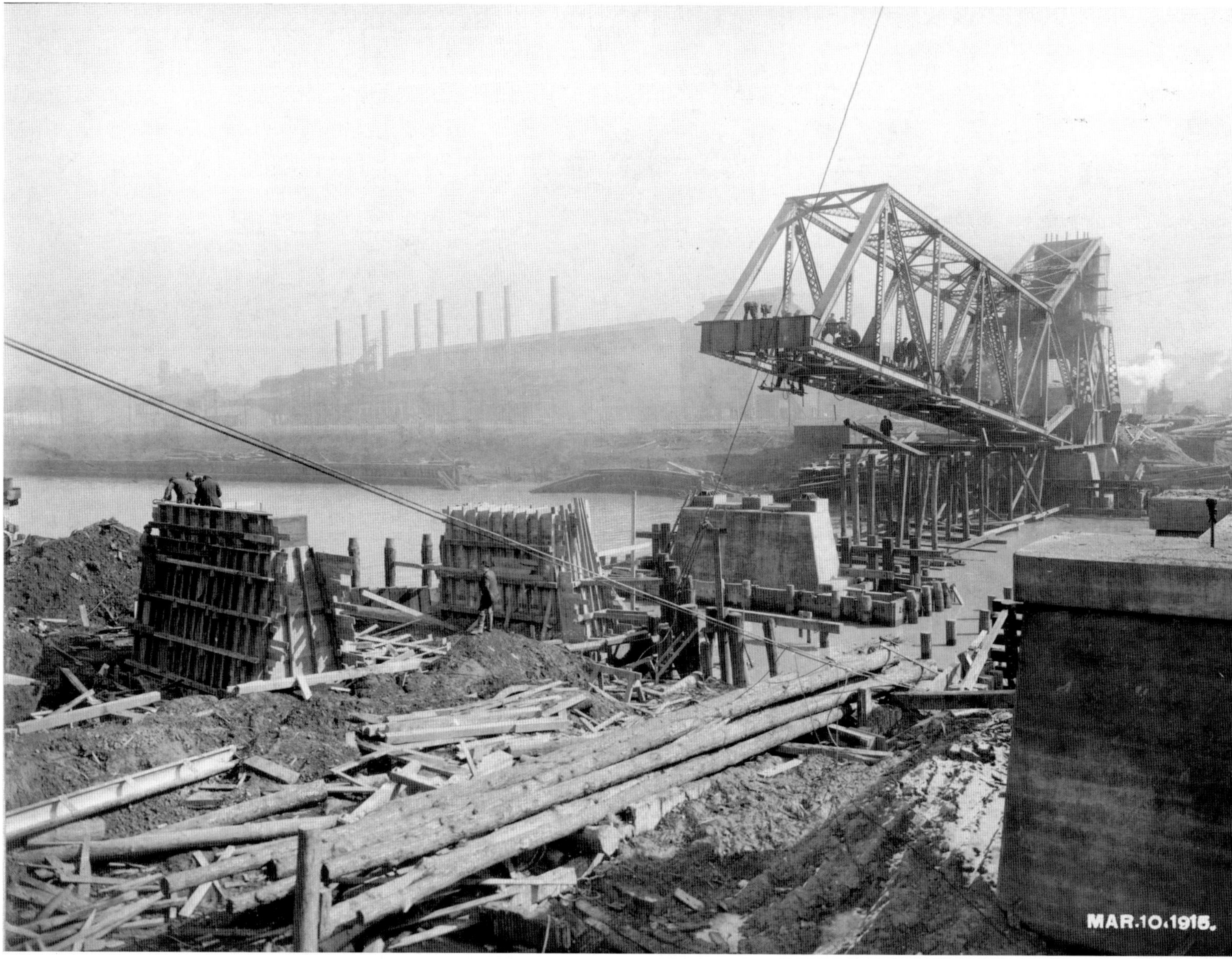

Looking northeast from the west side of the Cuyahoga River. The new bridge for the River Terminal Railway is a single-track Scherzer rolling lift bridge that opens for river navigation. In 1951 Republic Steel reinforced the bridge to accommodate heavier rail cars.

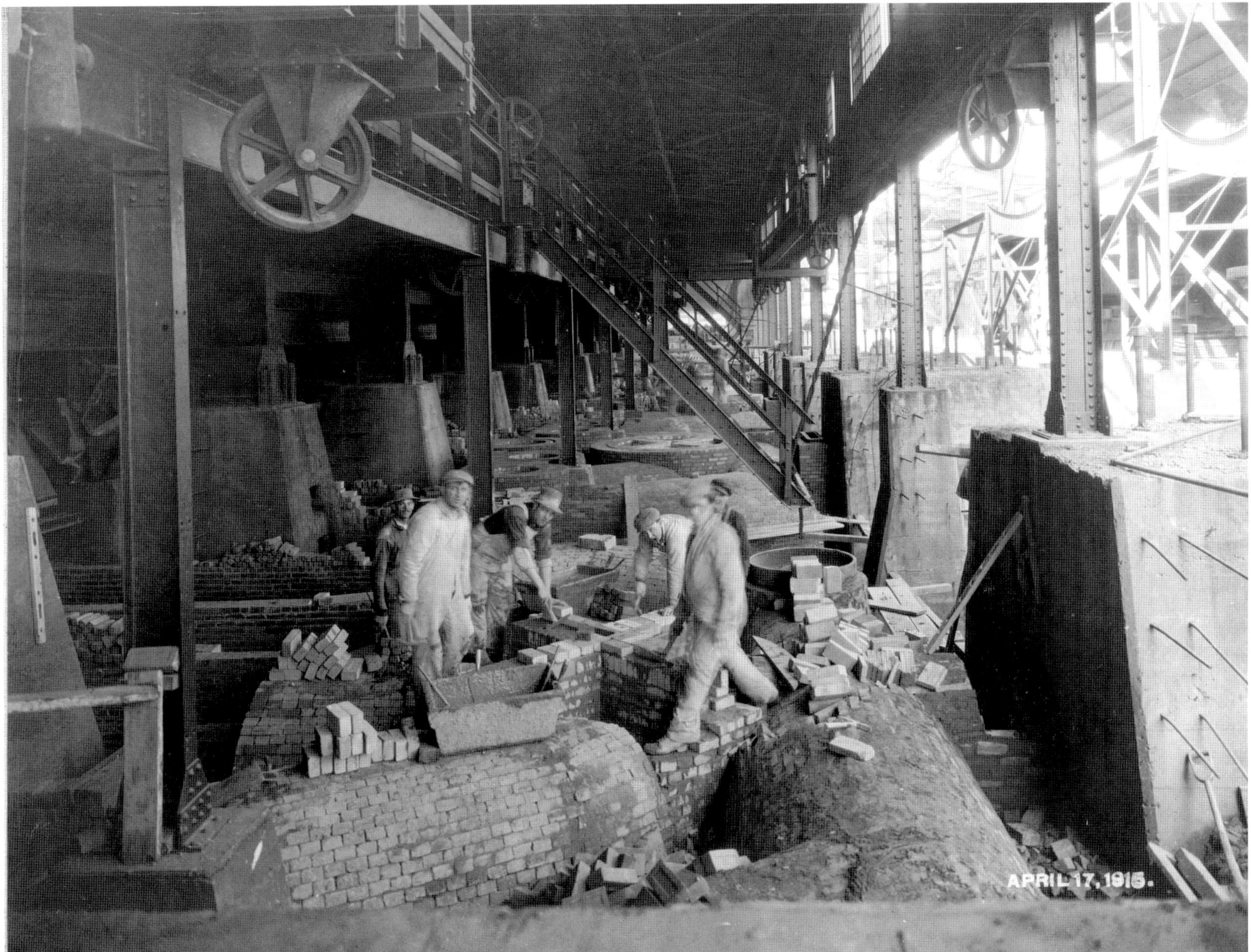

Inside the steel plant, looking north. Bricklayers build the soaking pit flue in this 1915 photo.

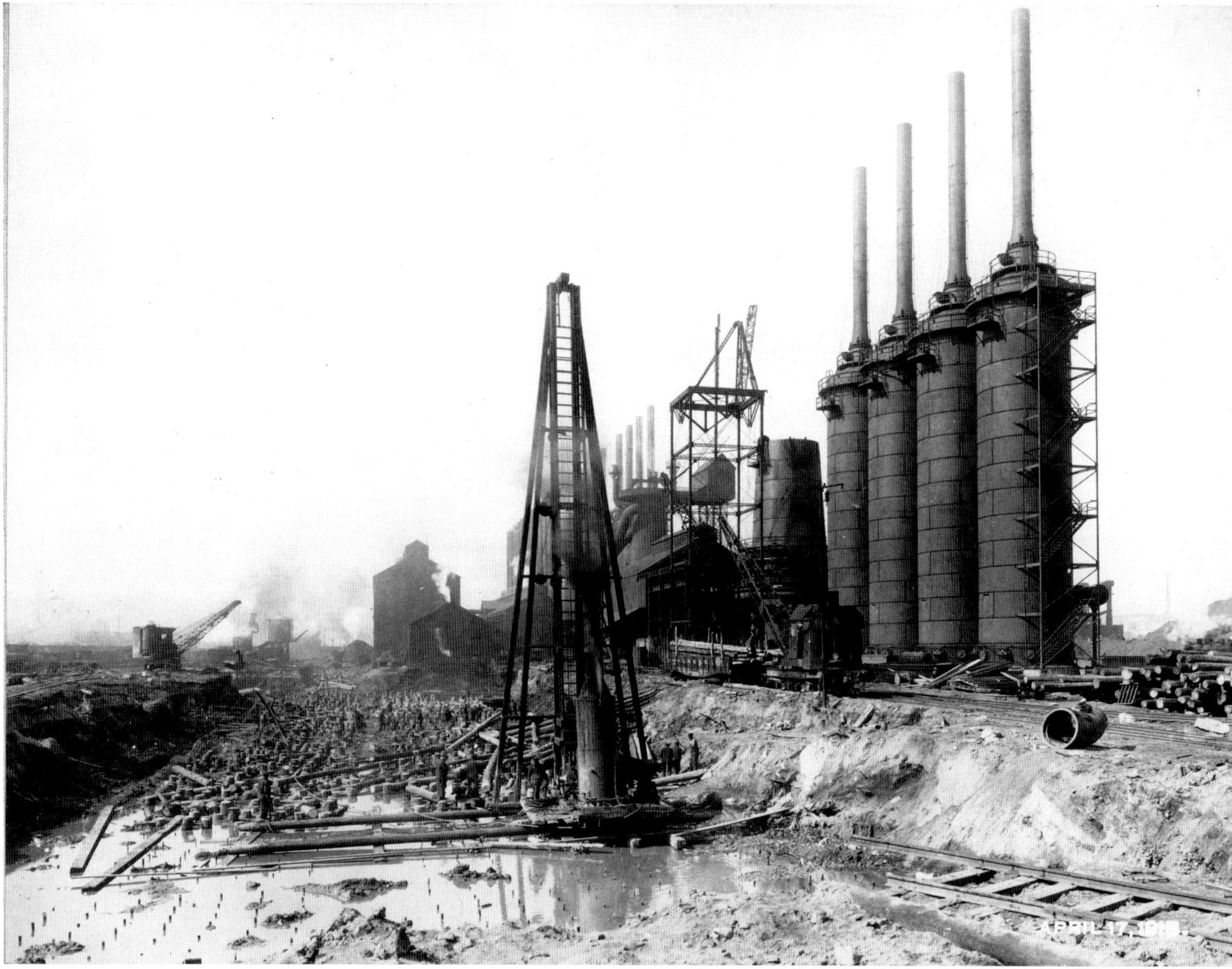

On the west side of the Cuyahoga River, looking north. The powerhouse foundation is in the center of this 1915 photo, and the stoves for Blast Furnace 3 are on the right. The rising shape of Blast Furnace 3 is just to the left of the four hot air stoves, which feed hot air into the blast furnace.

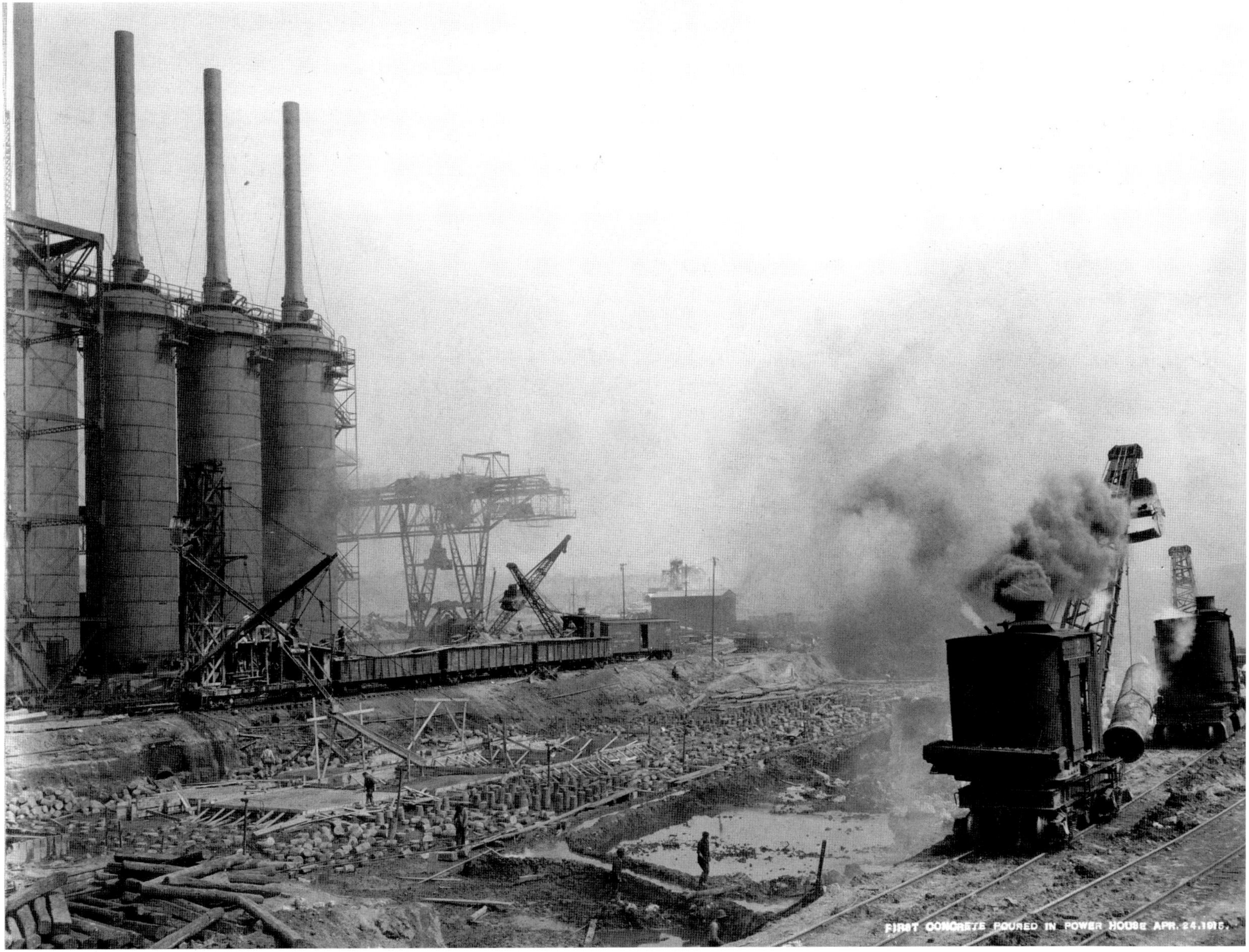

The photo says"First concrete poured in power house, Apr. 24, 1915." The four hot stoves from Blast Furnace 3 are on the left. Directly in front of the hot stoves is a concrete mixer on the tracks pouring concrete.

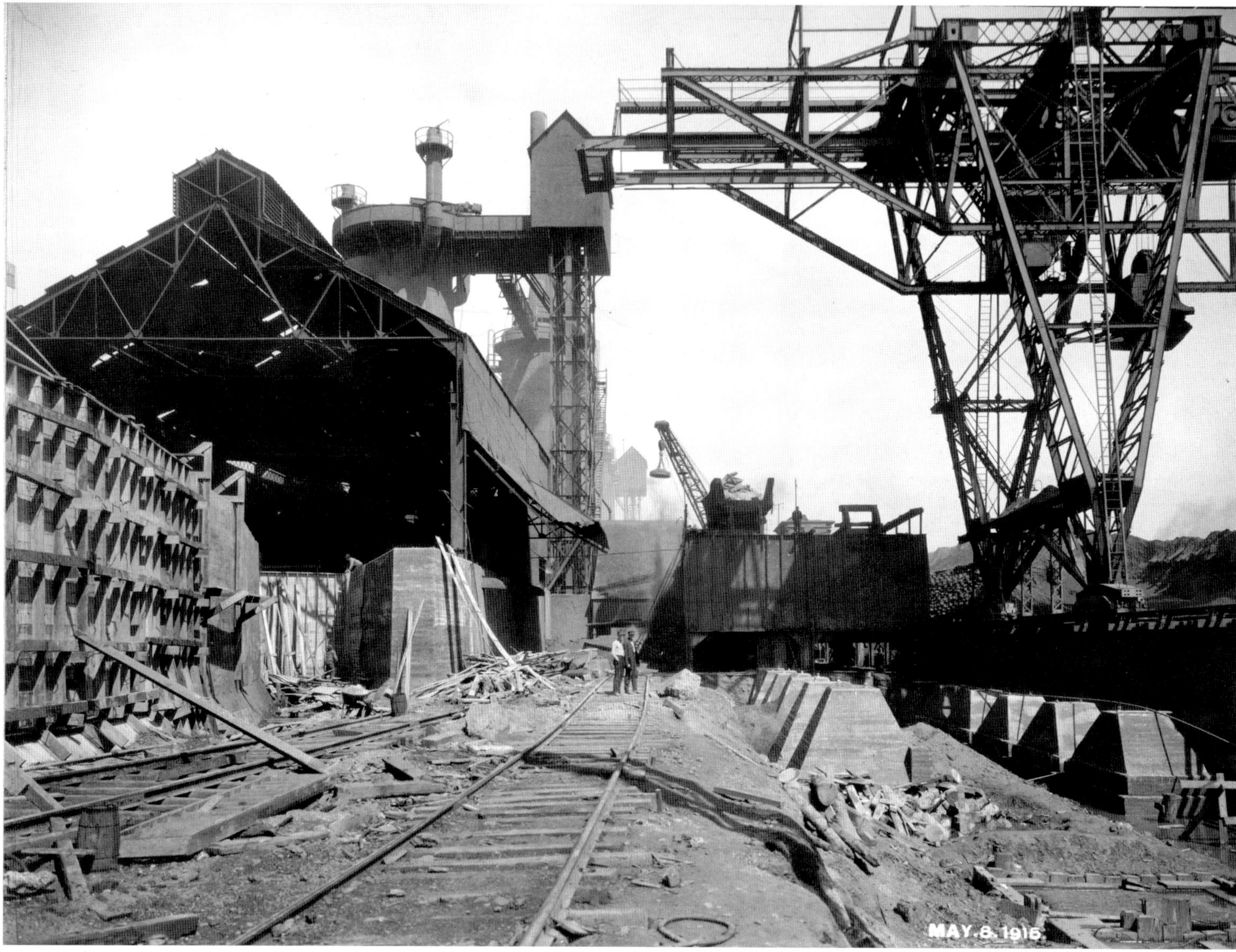

Looking north at Blast Furnace 2 in the background. Blast Furnace 3 is under construction on the left, and a Brownhoist ore bridge is on the right. Blast Furnace 2 is an older, vertical elevator type of furnace. Furnace workers transport raw materials in wheelbarrows up the elevator tower on the right of the furnace and across the bridge. They then dump the materials into the furnace by hand. This was a dangerous job, as rising fumes could overcome workers.

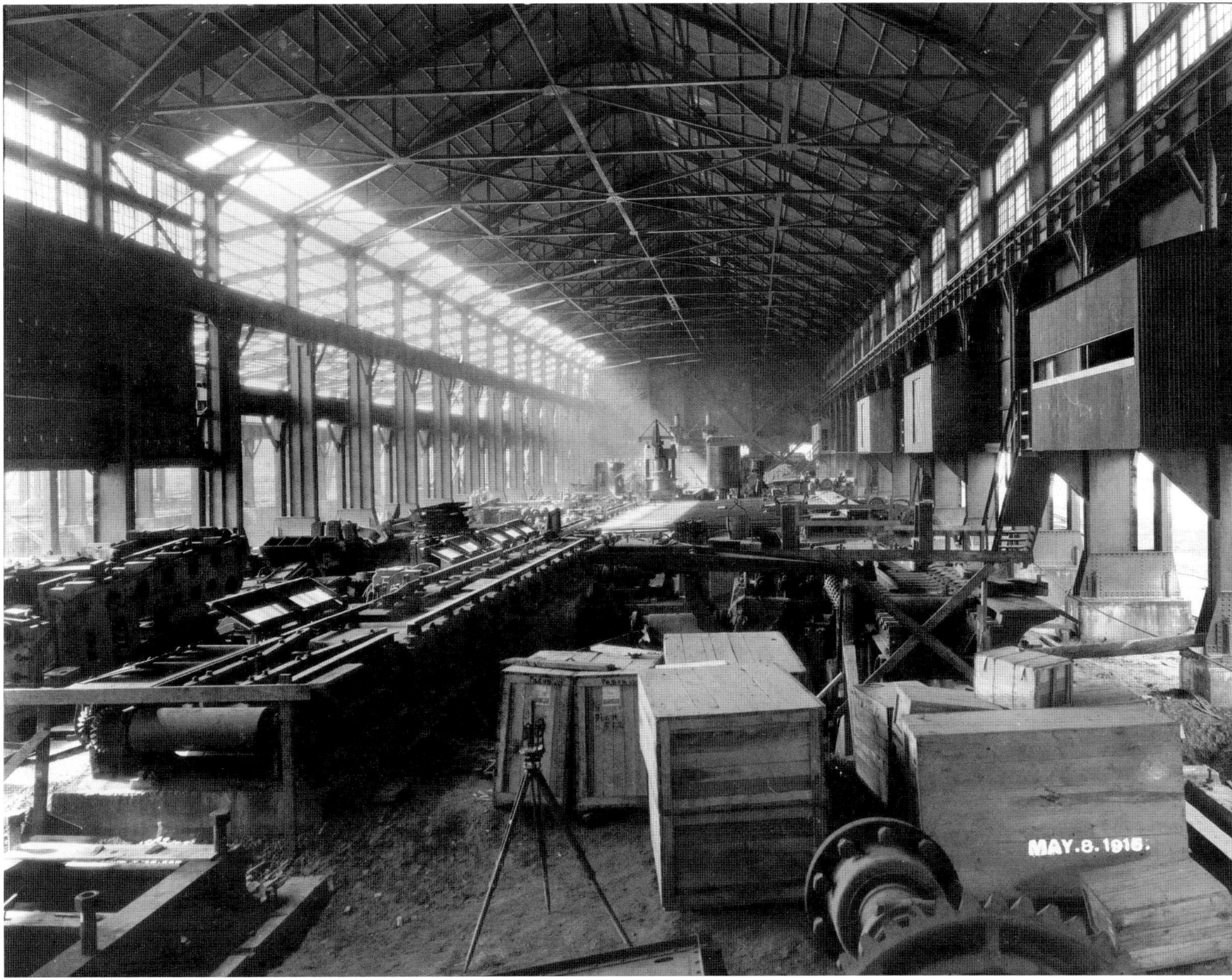

Inside the steel plant, looking south. In the center rear of this 1915 photograph, the 40-inch blooming mill is being installed.

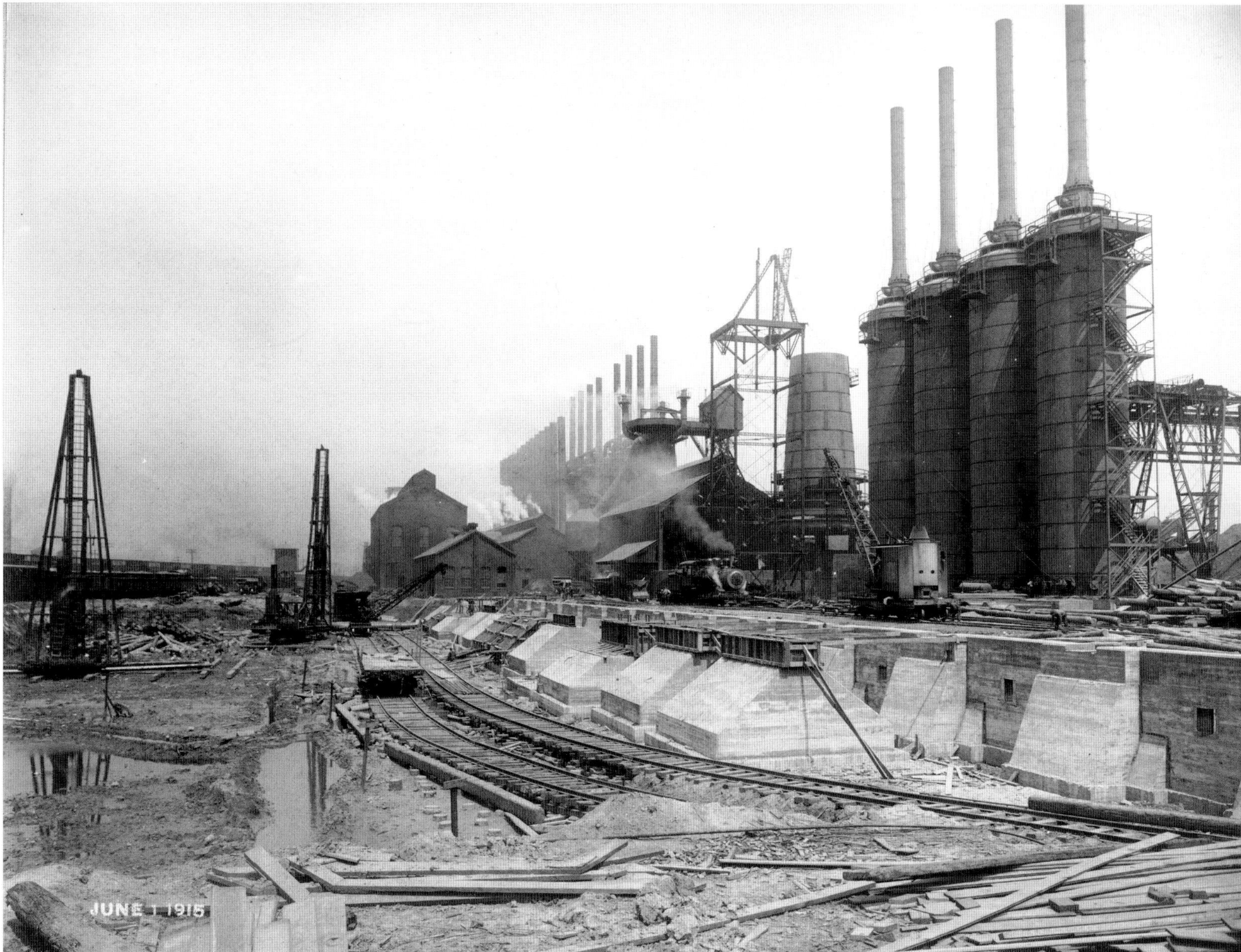

Looking north, with Blast Furnace 3 on the right and the powerhouse foundation on the left.

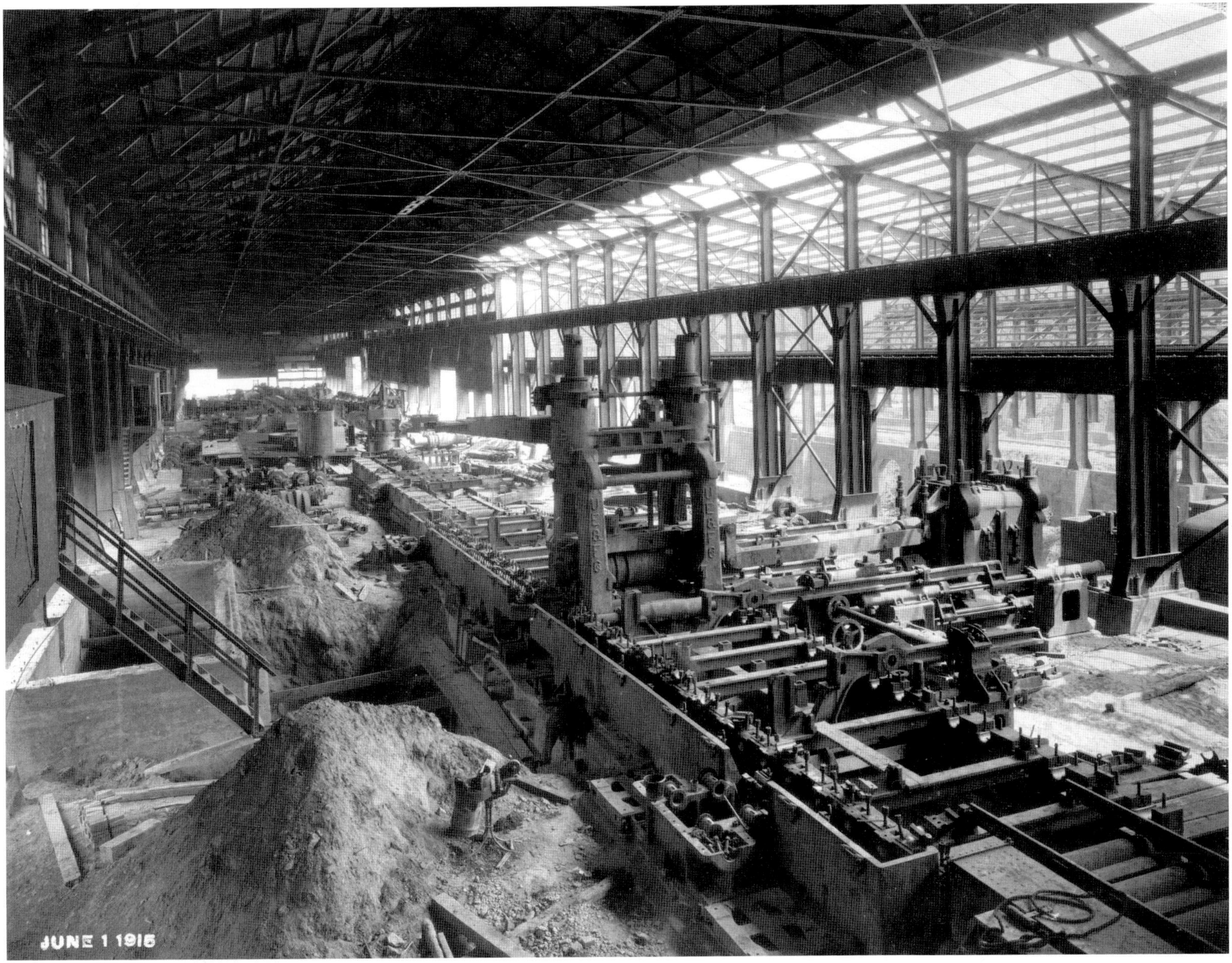

Inside the steel plant, looking north. In the foreground is the 40-inch blooming mill, which was built by the United Engineering and Foundry Company of Pittsburgh.

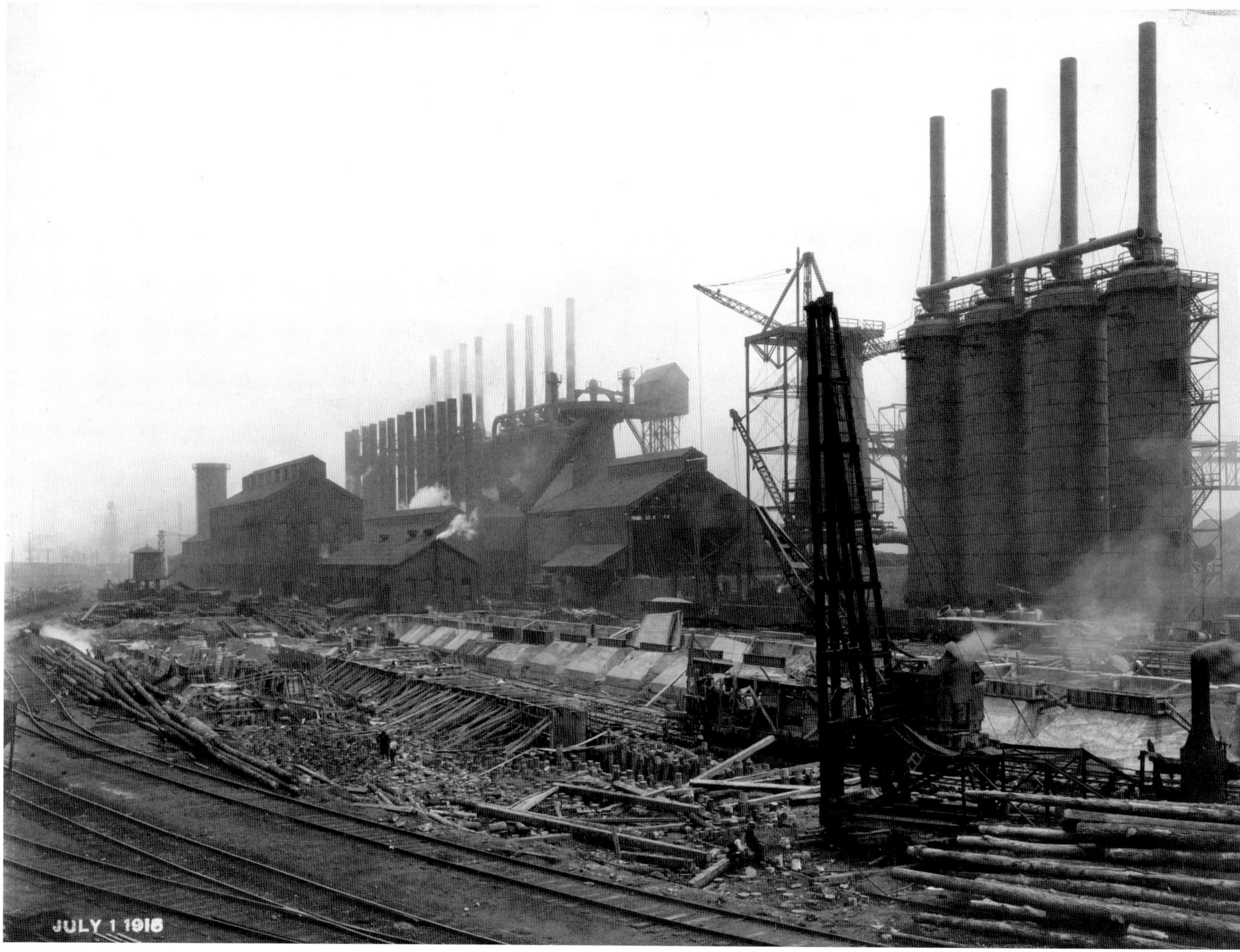

Looking northeast at Blast Furnace 3 on the right and Blast Furnace 2 on the left. In the foreground is the powerhouse foundation.

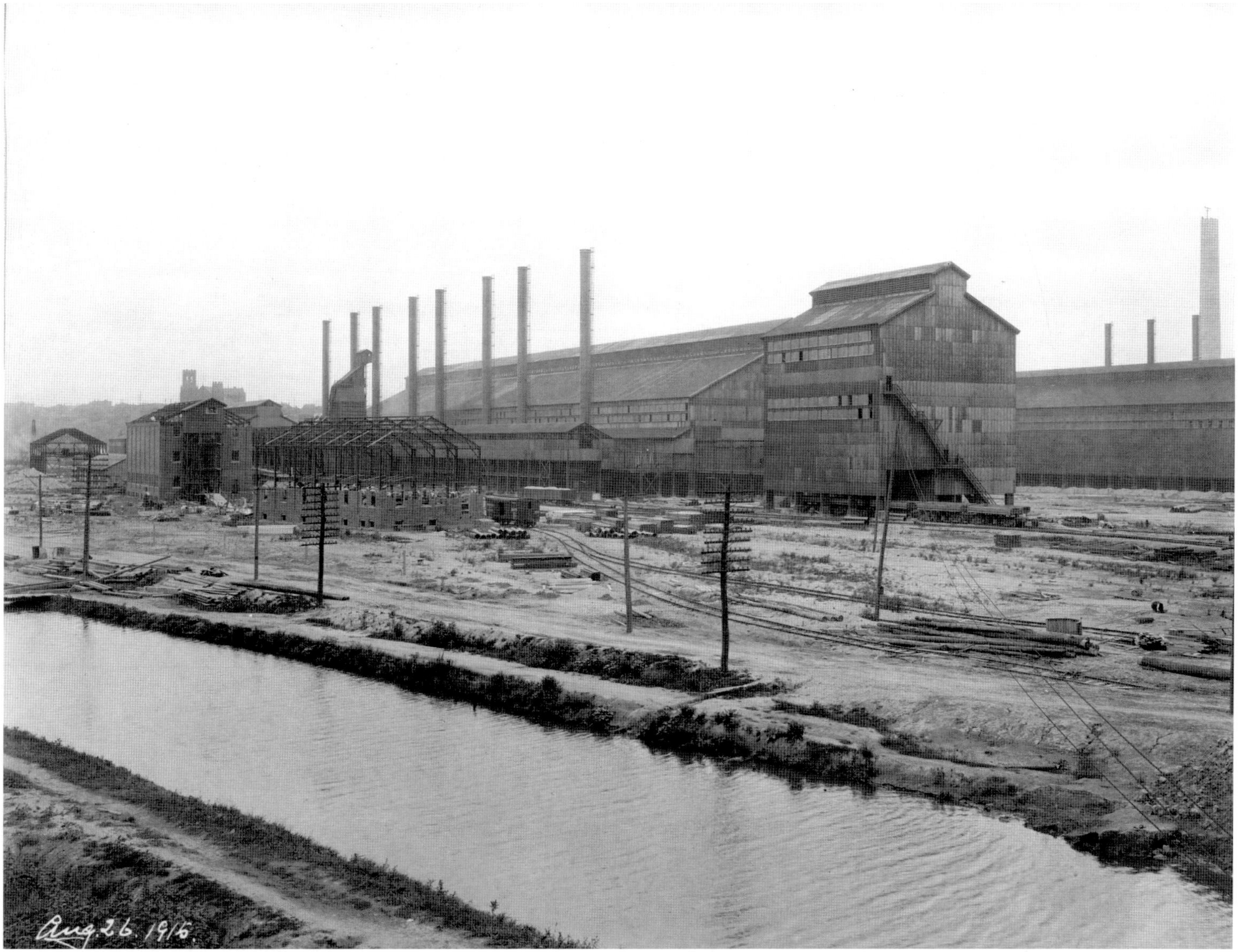

Looking northeast across the Ohio and Erie Canal at the steel plant in 1915. The mixer building is on the right with the open hearth furnace shop behind it. On the left is the storehouse, under construction, and the carpenter-pattern shop. St. Wenceslas is in the far background. A photo taken from this spot on a later date shows local children swimming in the Ohio and Erie Canal.

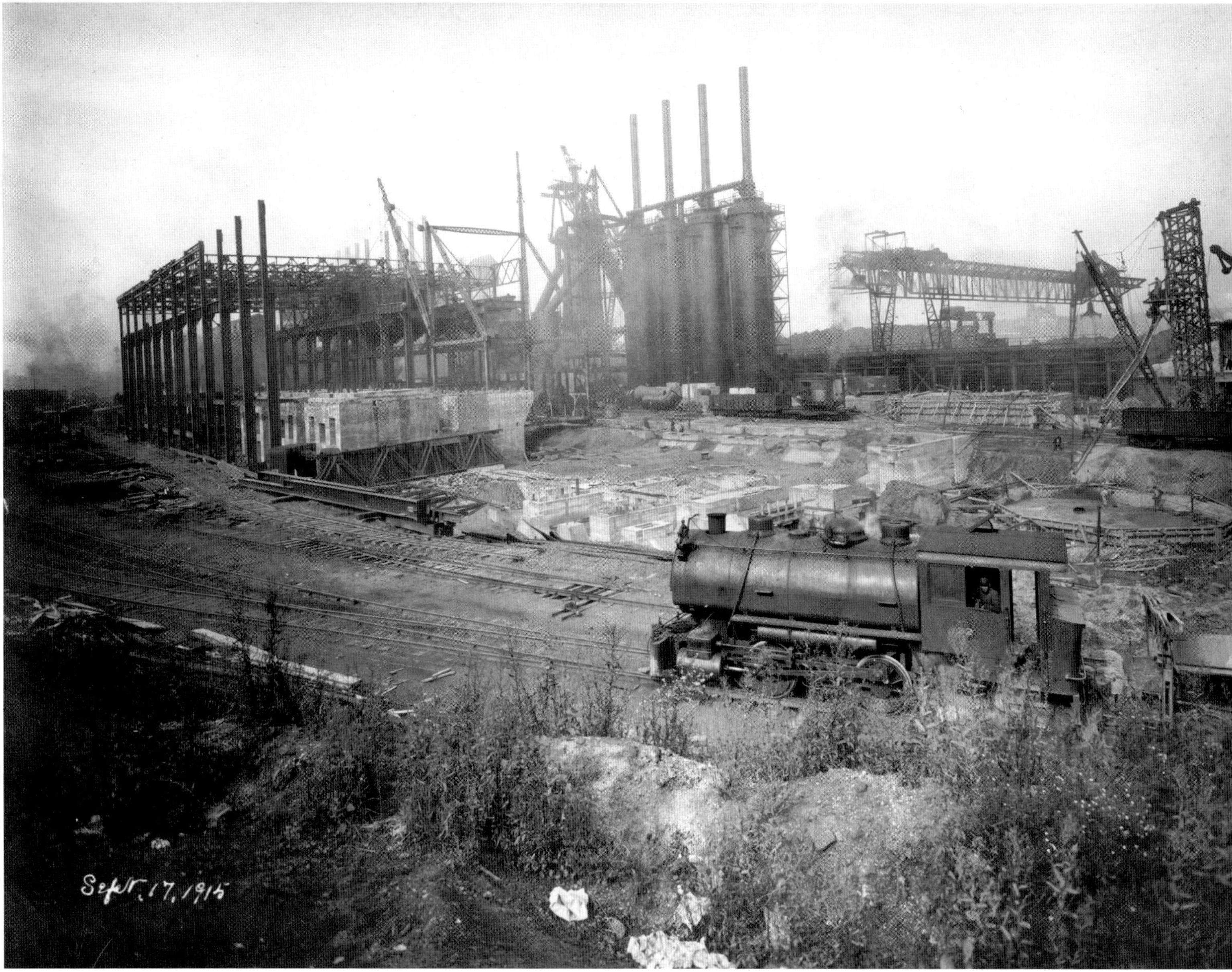

Looking northeast, Blast Furnace 3 is in the background, as the steel skeleton of the powerhouse rises to the left. The saddle tank switcher locomotive in the foreground appears to have stopped specifically to be in the picture.

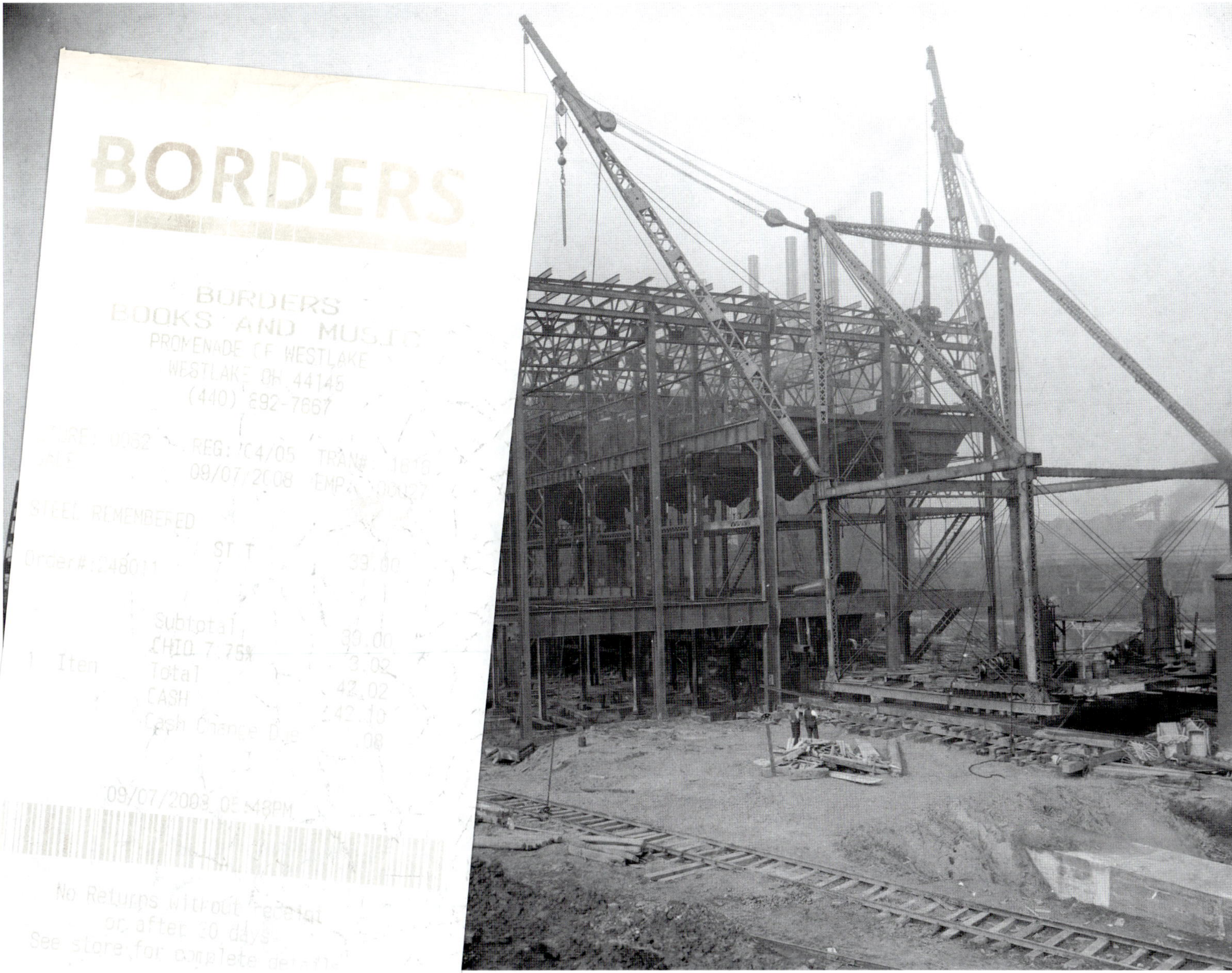

Blast [illegible] f the powerhouse completed. The skeletal structure on rails to the right appears to be a steam-powered [illegible]

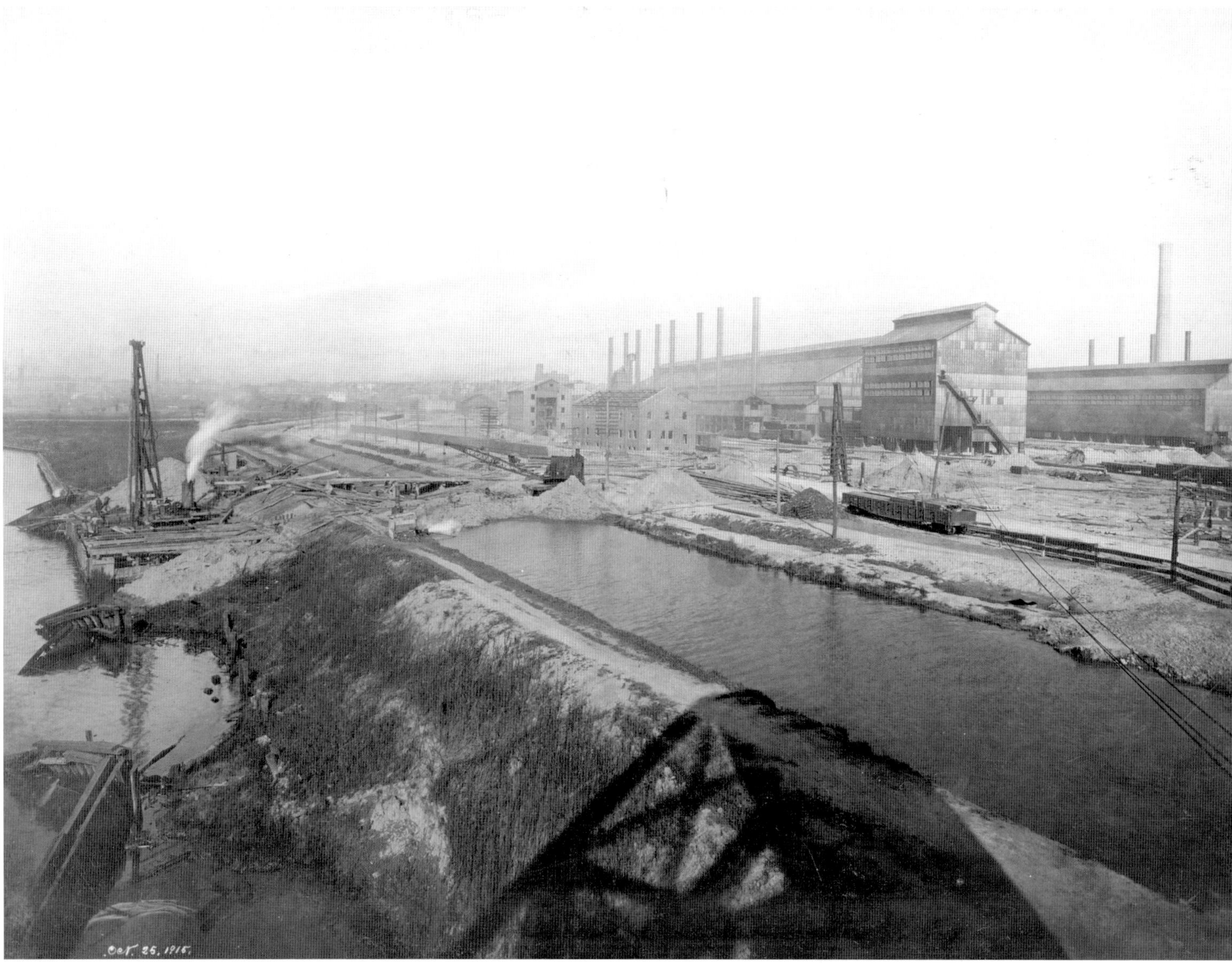

View from the Scherzer rolling lift bridge that crosses the Ohio and Erie Canal and the Cuyahoga River. Note the bridge shadow in the foreground. The steel plant complex is on the right, and most of the structures have been completed. The first pour of steel in this facility occurred January 20, 1916, about three months after this photo was taken. The rolling mills begin operation a few months later, capping off three years of construction.

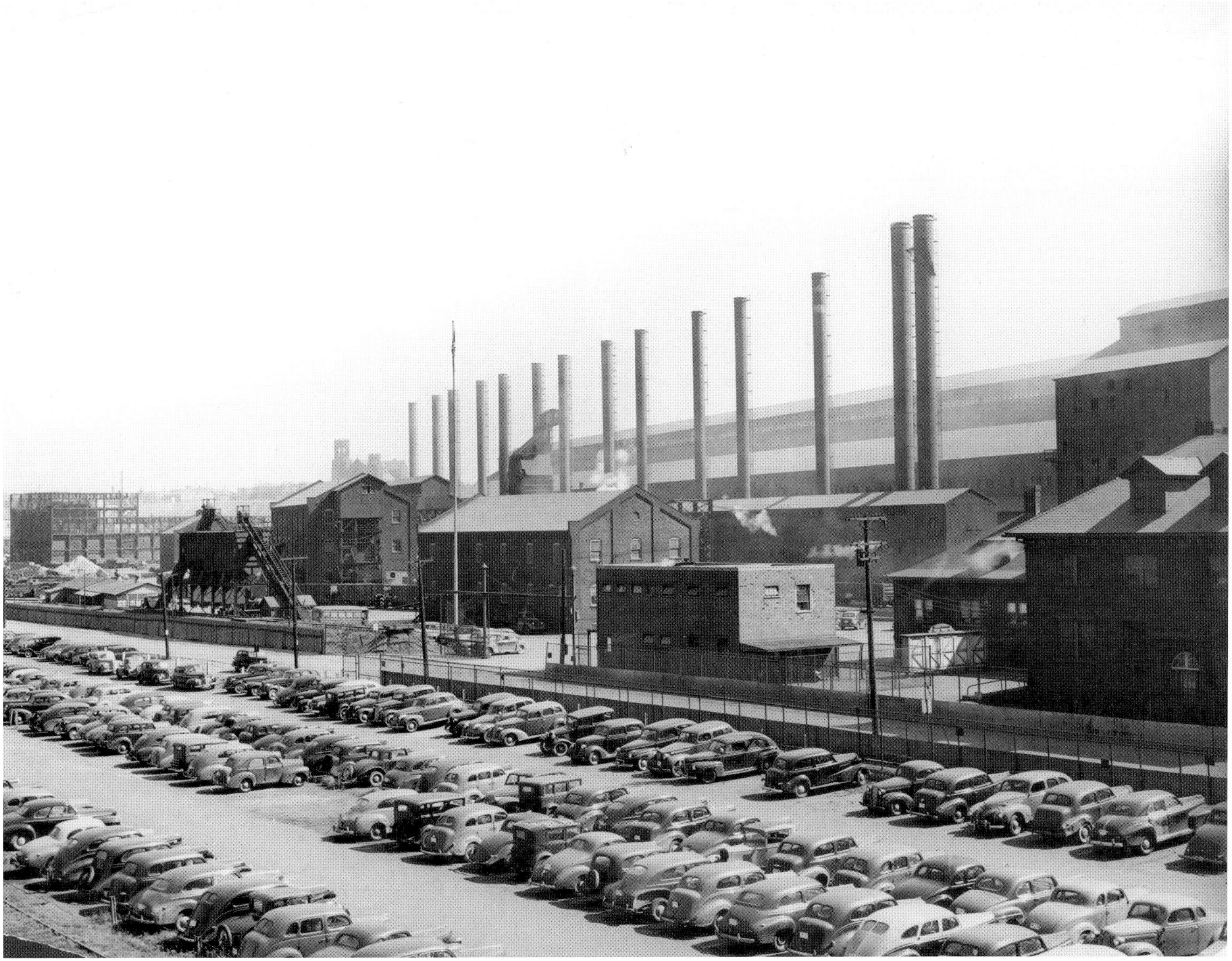

The Corrigan, McKinney plant in the late 1930s. Formerly part of the Ohio and Erie Canal, this area was paved over for employee parking, as shown here, and Independence Road. While a number of these buildings have since been torn down, some still remain, despite the mill's many owners, including Republic Steel, LTV Steel, International Steel Group, and today's Mittal Steel.

In January 1937, the Cleveland Press *ran a series of articles about the steelmaking process and the mills along the Cuyahoga River. The language in these articles is wonderfully descriptive, and conveys portions of the steelmaking process in an especially dramatic fashion. This article focuses on the shaping of steel in the strip mills of Otis Steel.*

"Man Masters Brute Metal," *The Cleveland Press,* January 5, 1937

Ira (Pop) Bryn, a hulking Welshman with powerful shoulders and big hands, is 54 years of age, and for 40 years has been working in steel mills.

Five days a week for eight hours at a stretch he is chief over a pair of roaring furnaces that spit out great blocks of glowing steel in an endless stream.

He knows his brutish charges down to their finest sparks. He can tell by the color of their fires almost exactly what temperature they register. To him the love of steel is as the love of wealth or power is to other men.

When he was but 14 years of age he began work in one of these thundering, smoking plants that grind out the basic ingredients of progress and war. His first job was as a scrap boy hired to pick up odd bits of steel. Today he is a "heater," a highly-skilled furnace man, in the Otis Steel Riverside mill.

Pop is typical of the thousands of Cleveland men who make their living through steel. He occupies a high post which few can achieve, but there are many veterans of 20 or more years in the mills.

Moving through the welter of the Otis plant, shocked and deafened by the jarring conflict of machines with hot steel, you find it hard to remember that humans keep these embattled giants under control. Yet there would be no such spectacular industry were it not for men like Pop.

At the receiving end of the Otis mill, sunk six or eight feet in a raised floor, are what men call "soaking pits." They are furnaces in pits into which the ingots, some shaped like the boles of big trees, some like huge cakes of ice, are lowered by a crane after they are removed from their molds.

For from two to five hours the ingots remain in the pits until they are within a few hundred degrees of the melting point. Once they emerge, dripping molten rivulets of slag and soft enough to be worked, they undergo a swift, brutal treatment that converts them into bars and sheets.

Vivid light as clear as transparent gold bursts free when the cover rolls back from a pit. It illuminates the men working in the background and envelopes the crane rolling up overhead to thrust down a pair of steel tongs.

Up comes the ingot like the rising sun. Gripped by the tongs it moves away through the mid-air, surrounded by the aura of its own brightness. It descends to a long table or bench on which rollers revolve in a continuous line. The crane drops it and retires.

In an elevated shed built on a steel bridge above the table, three men work a series of levers that control the next movements of the ingot. They stand facing a series of windows before which is the "40-inch blooming mill." Their shed is dark, but as the hot ingot propelled by rollers jolts by beneath their feet, their faces and the interior of the room glow momentarily with reflected redness.

A shock, a crash, a sharp concussion. Throwing off a fan of red-hot particles, the nine-ton ingot squeezes between two huge steel rollers, comes out thinner on the other side. Another crash. A metal arm strikes it and flips it over. Then back again through the blooming mill, again and again, reversing, squeezing out longer and thinner with each trip.

Working with perfectly synchronized movements, the three men in the shed control this activity. One operates the arms that flip the ingot, one the revolving table rollers that propel it back and forth, and the other the compressing rollers that exert pressure to "iron it out."

Suppose this particular ingot is one of those shaped like the trunk of a tree. If so, it weighs nearly 8000 pounds. It is pressed out gradually longer and thinner until it is several score feet in length, hissing and undulating like a tongue of flame as it slides back and forth through the mill. It is still red hot when it continues on to be eventually chopped into billets.

If, however, it starts out shaped like a cake of ice, it is destined for the "strip mill" and weighs nine tons. It is "reduced" in the blooming mill and then cut into longer and flatter cakes, and that is where Pop Bryn comes in.

Pop's twin charges are called continuous furnaces. They heat the cakes in one continuous operation, feeding them in at one side and discharging them re-heated on the other.

It is a noisy, lurid spot in the heart of the plant. The furnaces are like dragons, exhaling fire and black smoke and staring into the murk of the mill through doors resembling bloodshot eyes.

Glowing blocks slam out of the furnaces every few moments, crash down on to more revolving rollers and move away. A sharp swish, and mingled water and steam scrub the slab free of its coating of scale. With this process the strip mill operation begins.

A "continuous strip mill" is simply a series of presses that flatten the blocks into strips. From one press to the next the slabs go without pause, moving as fast as a man can walk until finally they writhe between the last rollers and emerge in long strips.

Some of the strips are cooled and cut into sheets which later are made into such things as automobile frames. Others intended for automobile bodies or washing machine shells, for example, are coiled and taken to an adjacent mill to be cold-rolled (rolled without being reheated) into still thinner strips.

Literally hundreds of finishing processes are employed to turn steel into thousands of sizes and shapes. Strip mill steel may be adapted to auto bodies, caskets, steel barrels, waste-baskets, tin cans and many other uses. It is expected to be an important item in construction of homes of the future.

Steel bars and rods are made into such things as tools and pipes and machine parts. Some of the steel leaves the plant in billets, big chunks to be "finished" by other firms.

When constructed eight years ago, the Otis strip mill was the world's largest. Others have out-ranked it in size since, but the Cuyahoga Valley will regain recognition with completion of the new Republic continuous strip mill next fall.

. . . Otis employs about 5000 in its two plants. There are about 6000 employees in four plants of the American Steel and Wire Company, which produces chiefly wire and wire products, and about 300 employees of the Carnegie Steel Corp., which makes pig iron.

The industry creates an army in Cleveland—an army of 18,000 workers manufacturing a significant portion of the nation's steel. Those workers with their families could constitute a city of perhaps 80,000 residents.

They live by the conversion of reddish rocks into a commodity of innumerable uses. Today it is ore heaped into great mounds alongside the docks. Tomorrow it is iron. The next day it is steel.

Who knows how long ago the fenders of new autos seen on the streets were nothing more than ore awaiting the fires of Cleveland's mills?

CHAPTER THREE
OTIS STEEL

Otis Steel, one of Cleveland's largest steel companies, had its start in 1852 when Charles Otis partnered with J. N. Ford and opened Ford and Otis, which later became Otis and Company.

Charles A. Otis was a native Ohioan, born in 1827. As a young child he moved to Cleveland with his parents, and while growing up he worked in his father's hardware stores and small iron forge. William Otis, Charles's father, had walked to Cleveland from Massachusetts and started as a mercantile dealer, then expanded into hardware, later adding iron production so that he could make his own hardware stock. He also was a banker and helped Charles get his start in business.

Charles Otis's education was sporadic at best, but he made up for it with perseverance and hard work. Initially a foundry, Ford and Otis cast railroad car axles and bar iron, then added a rolling mill in 1859. While this production continued during the Civil War, Otis sold his half of the company soon after the war ended and went abroad to investigate new steelmaking methods. One of those he learned about was the Siemens-Martin open hearth process. Impressed by the potential of the process to make high-quality steel in large quantities, Otis returned to Cleveland with a license to produce open hearth steel and founded the Otis Steel Company in 1874.

Otis Steel used the open hearth process exclusively and in fact was the first steel plant in America designed from the ground up (by noted steel mill engineers Samuel T. Wellman and Alexander Holley) to use the process. Initially the new plant, built on Cleveland's lakefront, produced steel through the acid open hearth process. In 1886 Wellman switched to basic open hearth steel, and again Otis was the first in the country to use this process to commercially produce steel. In 1889 Otis Steel was acquired by English investors after an English investment trust made a generous offer to Charles Otis and his partners. Otis sold the company to the British, but the company had an American president, George Bartol, and under his direction Otis Steel expanded its holdings in Cleveland.

After the turn of the century the company wanted to expand its main complex on the lakefront. There was little land available adjacent to its main plant, and the company championed the idea of using landfill to extend the shoreline of Lake Erie in order to build a giant steel plant with three blast furnaces and ore docks, allowing ore ships to unload directly at the blast furnaces. However, the City of Cleveland was reluctant to make such a major change to the lakefront. Other nearby port cities, including Lorain and Erie, Pennsylvania, made offers to Otis to move, but the company stayed in Cleveland. Looking elsewhere in the city limits, Otis located 330 acres along the Cuyahoga River, adjacent to the Cleveland Furnace Company, which had blast furnaces and a coke plant—the perfect supplier for a steel mill. Otis purchased the property in the river valley, and by 1914 its new mill, the Riverside plant, was going up. The original plant was renamed the Lakeside plant.

World War I proved to be highly successful for Otis Steel. Increased demand for steel in America and abroad caused the company to rapidly expand. Otis purchased the Cleveland Furnace Company to guarantee a steady supply of coke and pig iron without having to build its own blast furnaces. In 1919, after the war, a slump in the British economy caused the English investment trust to sell Otis Steel for needed cash. Cleveland banker John Sherwin organized an investment group in Cleveland to purchase the company and bring its ownership back to Cleveland.

In the postwar era the Riverside plant was expanded as more and wider rolling mills were added, especially when the auto industry was identified as a major buyer of steel. The Lakeside Plant focused more on large castings, wide steel plates, and basic steelmaking in the plant's older open hearth furnaces. The company also made a major addition with the selection of Elroy J. Kulas as the company president in 1925. Kulas was a respected businessman in Cleveland. Prior to World War I Kulas had been general sales manager of the Peerless Motor Car Company, a Cleveland-based maker of luxury cars. In addition to his duties as president of Otis, Kulas was president of the Midland Steel Products Company. But it was his background in automobiles that really helped Otis. Kulas secured automobile manufacturers as customers for Otis, making it a major supplier to the auto industry and boosting revenues in the late 1920s. Because of this, Otis added eight rolling mills with widths suitable for automobile manufacturing. A company newsletter in 1928 ran an excerpt of an NEA newspaper article about Kulas:

> Kulas looks like an easy-going, happy-go-lucky chap, of ready wit and good humor, slightly inclined to be portly, full of face with hair fast turning gray. But his stature belies his nature. He's dynamic, a quick thinker, a hard worker and keen psychologist. Kulas knows men and relies on them. . . . Having built up an organization of men in whom he has full faith, Kulas and his staff have gone out to "get" the automobile business for his two steel firms . . . there's nothing wishy-washy about this man. He's straightforward and undaunted. He has worked hard to get to his present position, but showed such skill and ingenuity as a salesman, executive, and student of mankind that he quickly rose to success.

In 1931 the Riverside facility added a new rolling mill, which for a time was the widest in the country. It was later widened in 1937. More open hearth furnaces were also added, as the directors of Otis Steel worked to make the Riverside plant an integrated steel producer.

But for its employees, Otis was more than just a steel producer.

The company worked hard to create a sense of loyalty and make workers feel like part of a larger family. It sponsored many social welfare programs and a variety of employee athletic teams, ranging from baseball to bowling. Some of the teams faced only other teams from Otis Steel, while others competed with industrial teams from Cleveland and other cities. Otis also sponsored musical groups, including an orchestra, a jazz band, and even a "hillbilly" band. These activities and others were covered each month in the employee magazine, *The Otis Sheet,* which often read more like a small-town newspaper than a corporate magazine. By 1929 Otis had more than 4,900 employees, which meant the mill was bigger than many small towns. *The Otis Sheet* often featured safety notices, inspirational messages from E. J. Kulas, articles about company history, photographs of newly married employees, photographs of employees' children, sports scores from the various Otis department teams, and all sorts of departmental gossip.

A good example of this "small-town feel" is evident in the January 1935 issue of *The Otis Sheet.* That issue featured a page-long obituary for Russel L. Davis, the Otis general superintendent who had died of a heart attack on December 12, 1934; an article about a number of promotions for Otis workers; a photo of Carol Louise Pekar, the one-year-old daughter of Otis employee Randall Pekar; an article about the wedding of John Kelly, from the sheet mill pickle department, who married Miss Lenore Wichman; a photo and article about Johnnie Lawson, an African American employee in the shipping department who had recently celebrated his twenty-fifth wedding anniversary with his wife, Maggie; several short obituaries for Otis employees or family members of Otis employees; team standings for the Otis Steel bowling league (with the Blanking Department and the Sheet Mill Warehouse tied for first place, though the Sheet Mill Warehouse had the top two bowlers); an article about the Otis band, which played at a charity benefit in Public Auditorium; and an amazing amount of employee gossip and bad jokes, including:

> Bill Hiskey knows a guy who's all wrapped up in a beef-stew romance. She's always beefin,' and he's generally stewed.
>
> Dolly Hines is training again. Does his rope skipping and shadow boxing for four hours, and then punishes beer for another four hours.
>
> Joe McMahon enjoys his little joke—so much so that he is beginning to tell them on himself. The boys thought it quite sporting of him when he related the following: "I had a few on my way home. Well, maybe more than a few. The next day the wife told me that I marched straight up to the ice box and ordered a straight and a beer." Joe is still waiting for that order to be filled. Or perhaps waiting for the 60 days for talking to himself.
>
> Maybe you have noticed Nick Zema and Joe Specht of the weld shop recently. Their facial features seem to be undergoing a change. Some of the boys feel it's just a bit of asbestos for protection of the upper lip that they have pasted there.

By the mid-1930s Otis Steel was beginning to show its age. While surviving the Great Depression, the company was not updating its equipment, which was steadily degenerating. Cleveland financier Cyrus Eaton had considered adding Otis Steel to his burgeoning Republic Steel empire in 1930, but outside consultants felt the plant wasn't modern enough and would cost too much to bring up to date. Eaton looked elsewhere, and in 1935 Republic acquired Corrigan, McKinney Steel, directly across the Cuyahoga River from Otis. Despite all that Otis Steel continued in production until 1942, when Pittsburgh-based Jones and Laughlin corporation, seeking a plant on the Great Lakes, purchased the company.

Initially named the Otis Works, the plant was later renamed the Cleveland Works, and was one of three main steelmaking facilities operated by Jones and Laughlin for the rest of the company's existence (the Otis Lakeside mill was closed and dismantled after World War II ended). In the postwar era, Jones and Laughlin expanded the facilities, and invested hundreds of millions of dollars into new equipment and buildings, making the Cleveland Works one of the most modern mills in the country. With new equipment, production levels soared in the 1960s and 1970s, as Cleveland Works employees constantly set production records. By this time Jones and Laughlin was no longer an independent company. The company had been acquired by Dallas-based conglomerate the Ling-Temco-Vought Corporation, also known as LTV. LTV subsidiaries made everything from airplanes to steel, from sporting goods to carpeting. To say it was an unwieldy combination is to be charitable. The company was on an endless cycle of mergers, acquisitions, and sales.

In 1978 LTV merged with the parent corporation of Youngstown Sheet and Tube and consolidated that company with Jones and Laughlin, making LTV the third largest steel company in America. While the size of the company was impressive, it could not hide the fact that the steel industry was hurting. Demand was dropping, imports were rising, formerly modern equipment was becoming obsolete, and steel companies were bleeding red ink. Facing an uncertain future, LTV Steel sought out its competitor Republic Steel for a proposed merger. In 1984 a merger was consummated and two years later the new LTV Steel Corporation, the second biggest steelmaker in America, slid into bankruptcy for the first time. The company reorganized, selling off many of its Republic Steel facilities, and eventually emerged from bankruptcy. In Cleveland the merger between LTV-owned Jones and Laughlin and Republic Steel finally united the two mills on the banks of the Cuyahoga into one complex, until LTV entered its final bankruptcy in 2000. When the newly formed International Steel Group purchased LTV's assets, the Cleveland Works resumed operations, but the entire complex was not fully reopened, as much of the old Otis Steel portion on the west side of the Cuyahoga River remained shuttered. In 2005 most of the Otis portion of the complex was scrapped; machinery was sent to China and buildings were torn down. The site today is home to the Steelyard Commons shopping center, and sales are made where steel was once made.

. . .

When Deb Kavulich at LTV showed the staff from the WRHS the boxes of negatives and photographs that make up the LTV Steel Collection, she also pointed out a large box off in the corner, which contained Otis Steel photographs. The staff from WRHS was a bit surprised, because few if any Otis Steel photographs existed in Cleveland. Neither WRHS nor any of the other major archives in the city had more than a handful of Otis Steel photographs. It was assumed that any photographs of the mill had fallen into the hands of Jones and Laughlin and had been sent to Pittsburgh, where they were eventually discarded or destroyed. But there in the basement of LTV's Cleveland Works building was a large box containing hundreds of photographs from Otis Steel, dating back to the early days of the Riverside mill. There were also a number of pictures of the historic Lakeside mill, but only interior shots. Unfortunately there were no photos shot by Margaret Bourke-White, who had used her ethereal photos of Otis to catapult herself into the front ranks of photojournalism. The photos in the box were mostly taken by staff photographers, and used in Otis Steel's in-house magazine, *The Otis Sheet*. The photos depicted machinery, giant castings from the Lakeside mill, and workers on the job. In one fell swoop, a long-forgotten company was brought back to life by a box of photographs.

The photographs in this chapter represent a cross-section of the Otis Steel photographs. They include photos of various mill processes and wonderful photographs of work crews at the mill in the mid-1930s. Only a few of the Otis photographs are credited to a photographer; when available, the photographer's name is noted in the caption. Many of these photos were used in *The Otis Sheet,* where Frank Cerness is credited as photographer. It is highly likely that he took the bulk of these photographs.

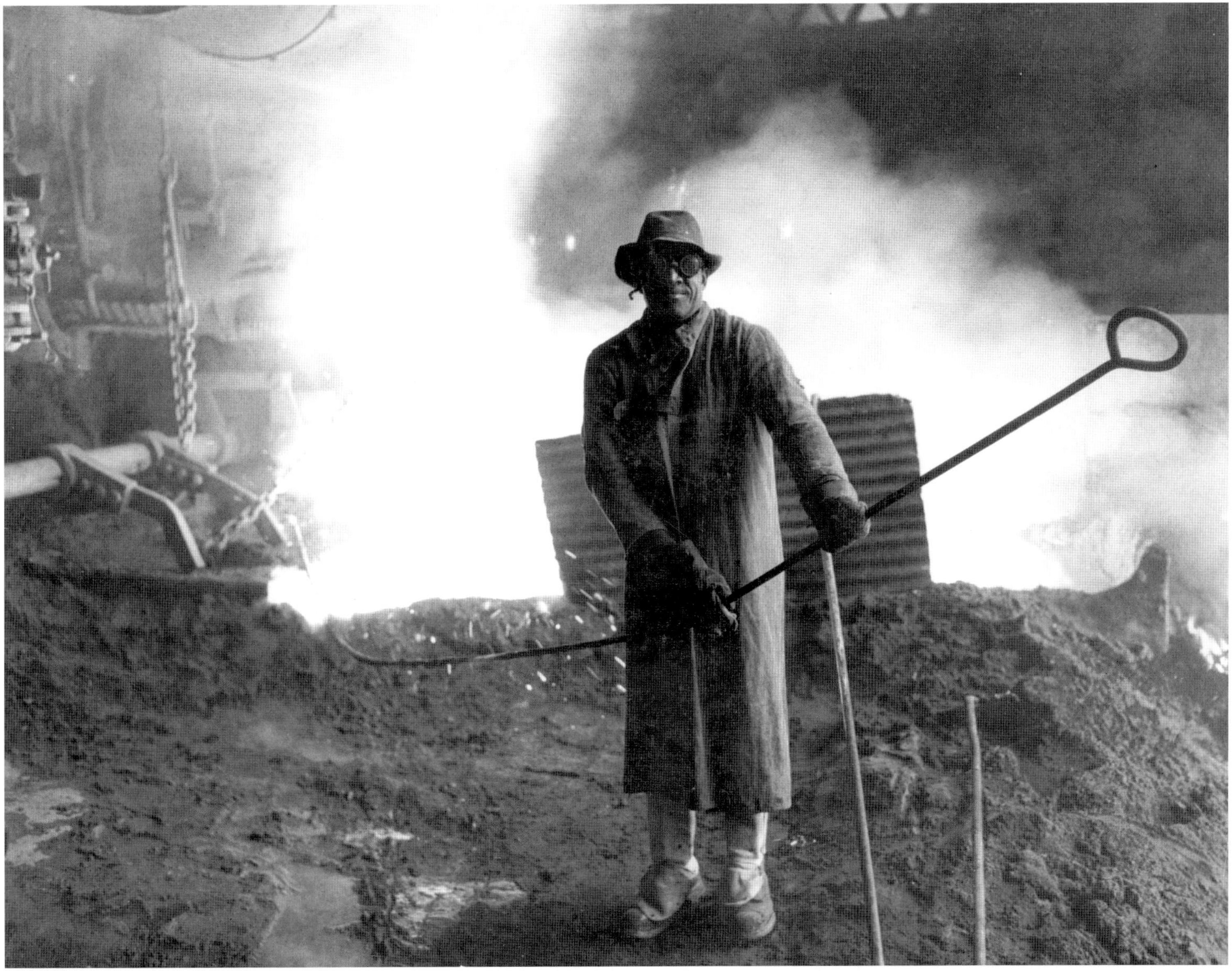

Photo of a worker in the casting house of the blast furnace. With one of the most dangerous jobs of the mill, the worker does have some safety clothes, though protection from a hot metal splash or explosion is dubious at best. Photo by Edward L. Feldman, Cleveland Heights.

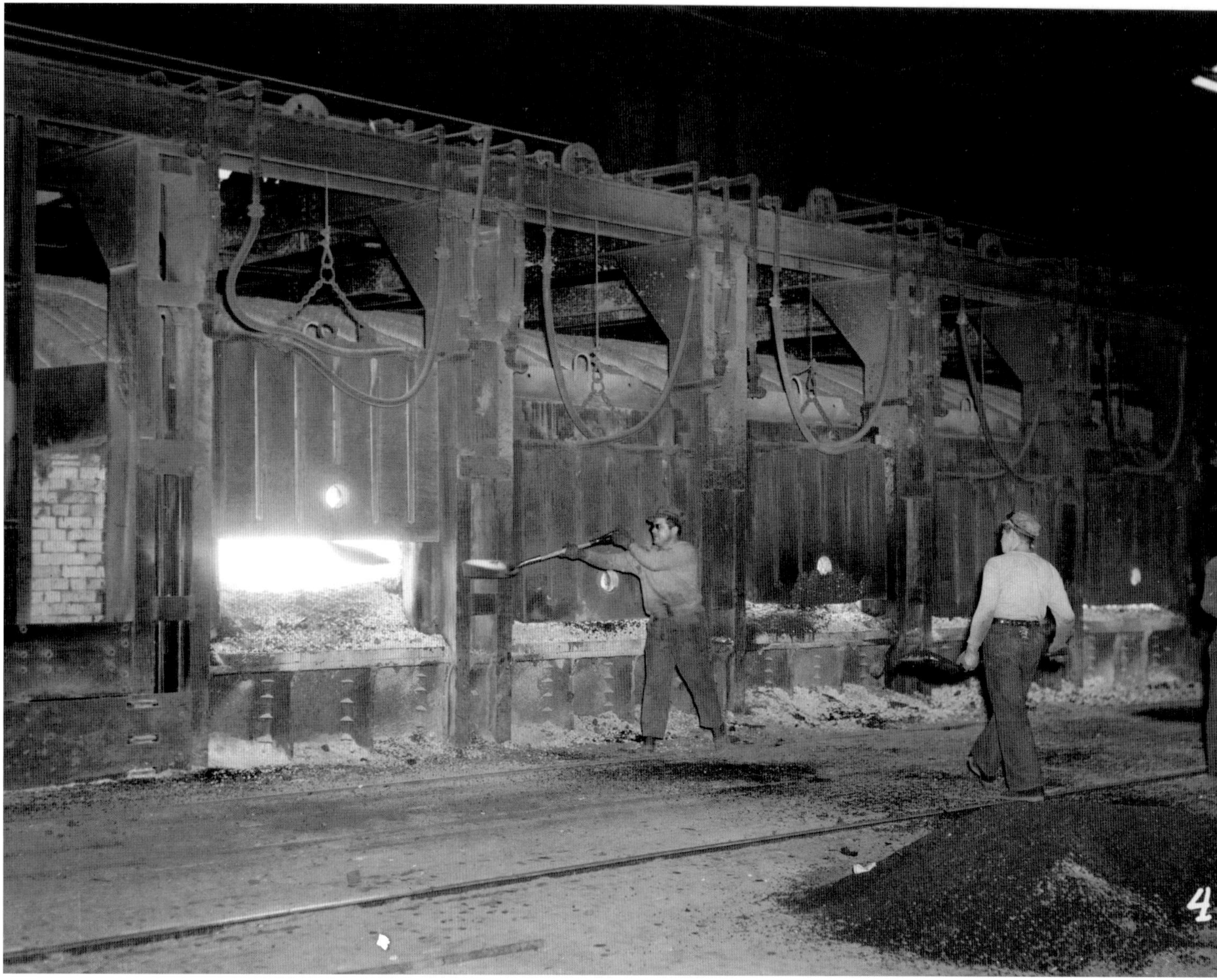

Open hearth workers shovel into the furnace what is most likely dolomite, a refractory material that helped protect the walls of the furnace from the molten metal within. The workers had to fling the material in such a way that it would land against the interior walls of the oven, often requiring them to get very close to the open port of the 3,000-degree furnace.

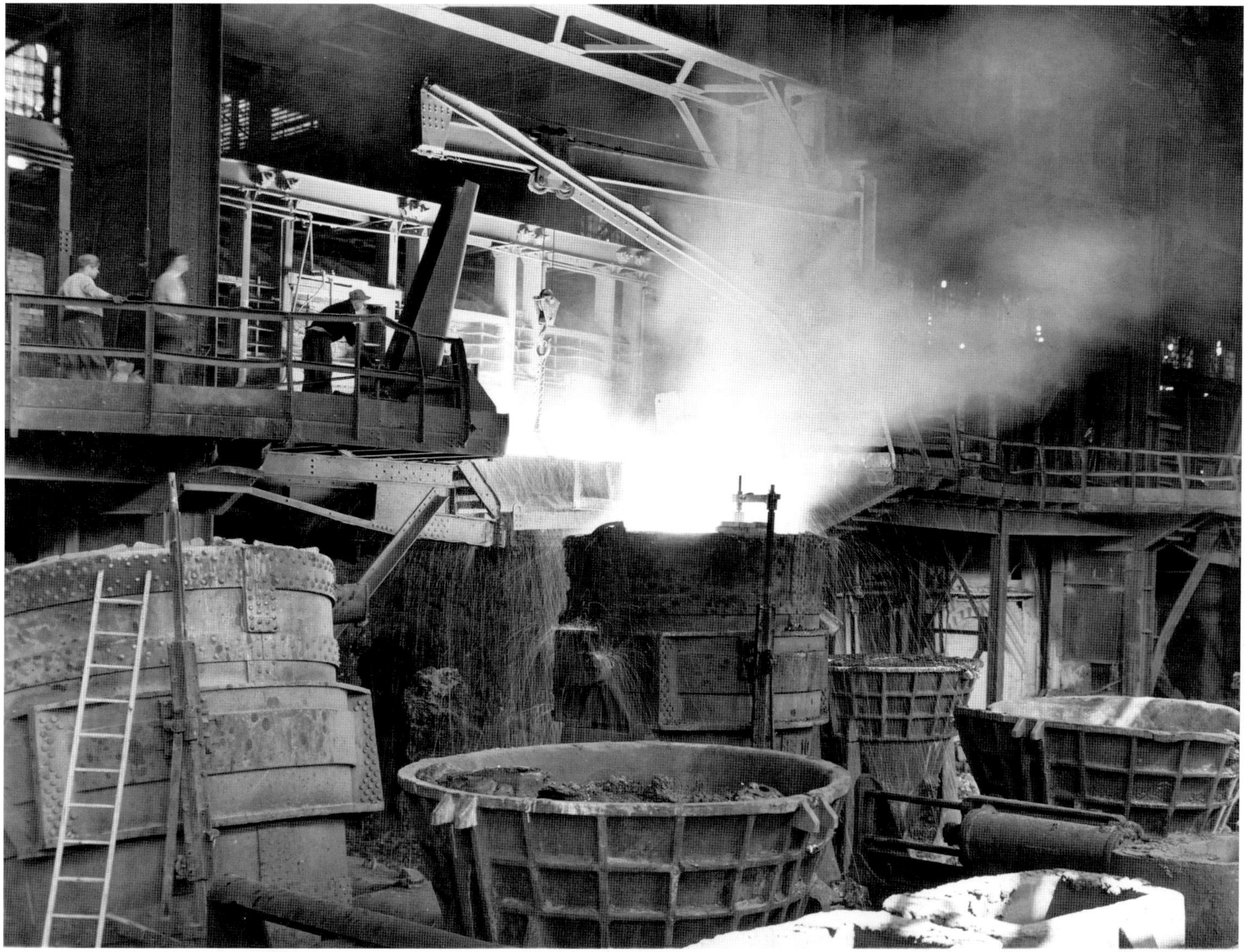

After the melter and first helper judge the molten steel in the open hearth to be the required quality, they tap the furnace and the molten steel pours into a waiting ladle. Open hearth workers would often fling bags of alloy materials into the ladle to change the composition of the steel per a customer's need. This was a dangerous part of steelmaking, as the platforms overlooking the ladles often did not have railings or safety equipment. Steelworkers could, and did, fall into the ladles of hot metal or into the pits where ladles sat and where hot slag or jagged metal pieces often littered the floor.

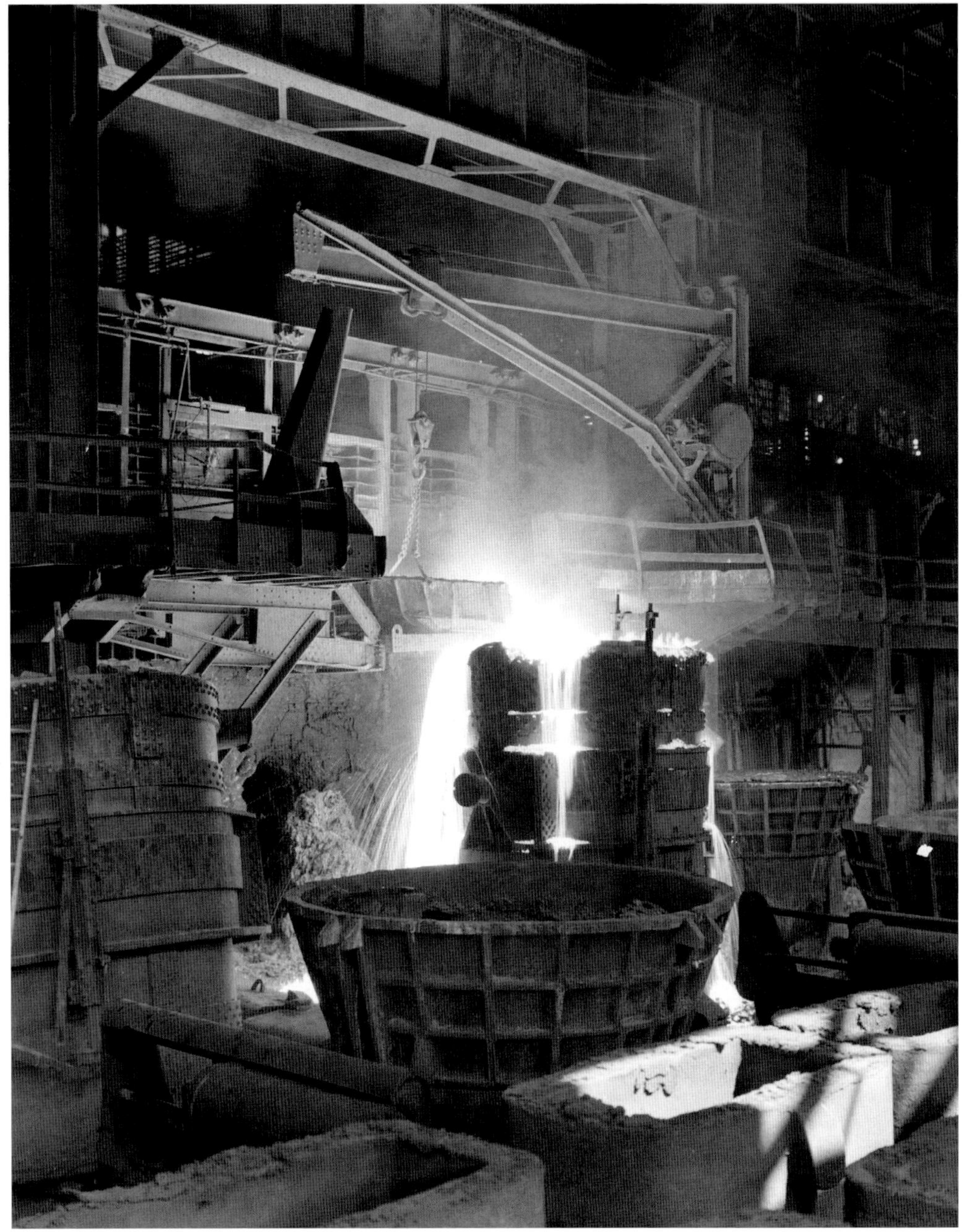

An open hearth tapping, dramatically illuminated by the light of the molten metal in the ladle and by sunlight streaming through a few windows high above. Open hearth shops were often not well lit or even heated. It was not unusual in some mills for workers to get frostbite in the winter, even though they were working within a few feet of furnaces containing molten metal.

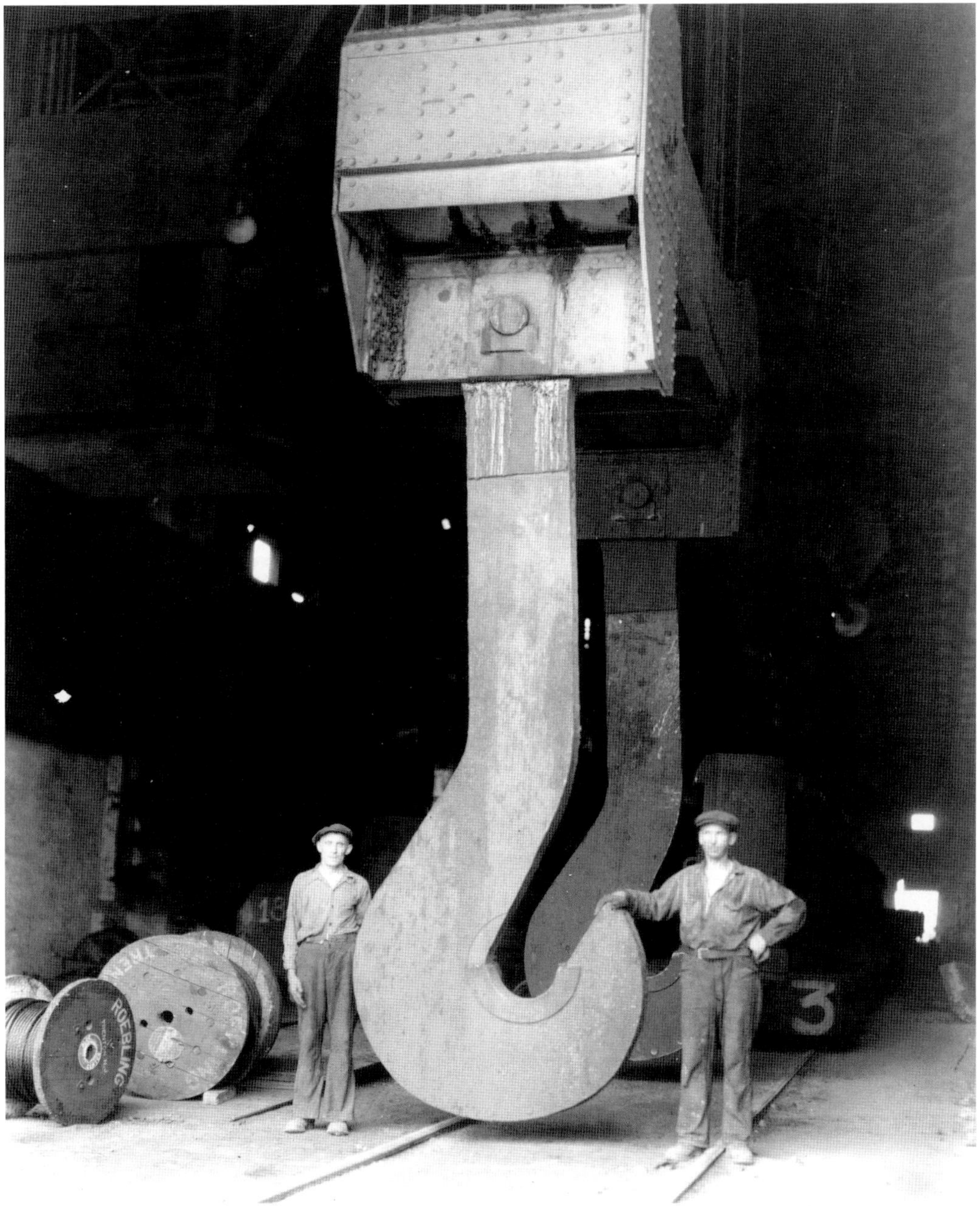

The mill was indeed a giant's playground, as everything was larger than life. Here an open hearth ladle crane dwarfs two workers, listed as H. Grendzinski and Steven Roth on the back of the photo.

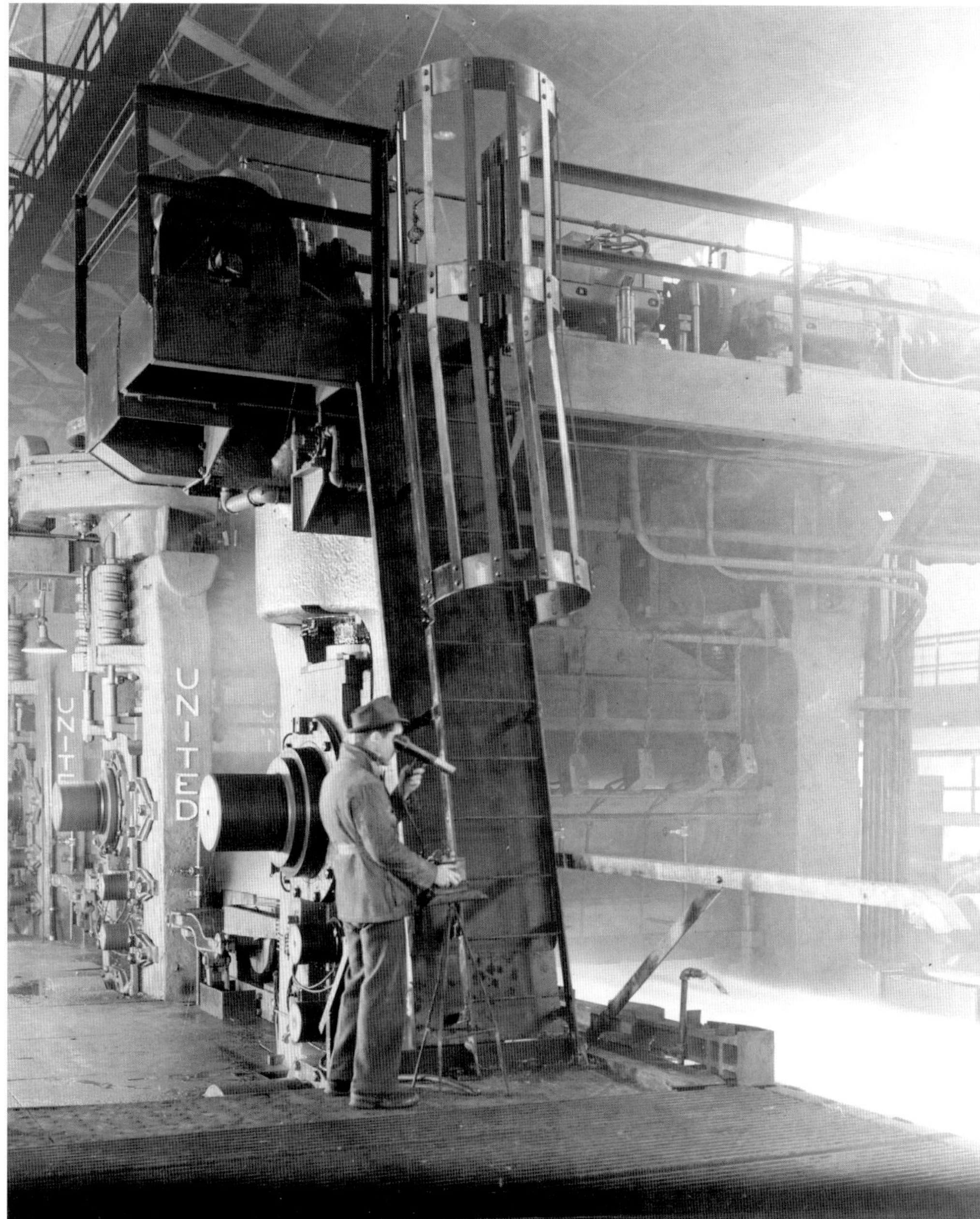

This worker watches as giant rolls squeeze a bloom into long, thin sheets of steel.

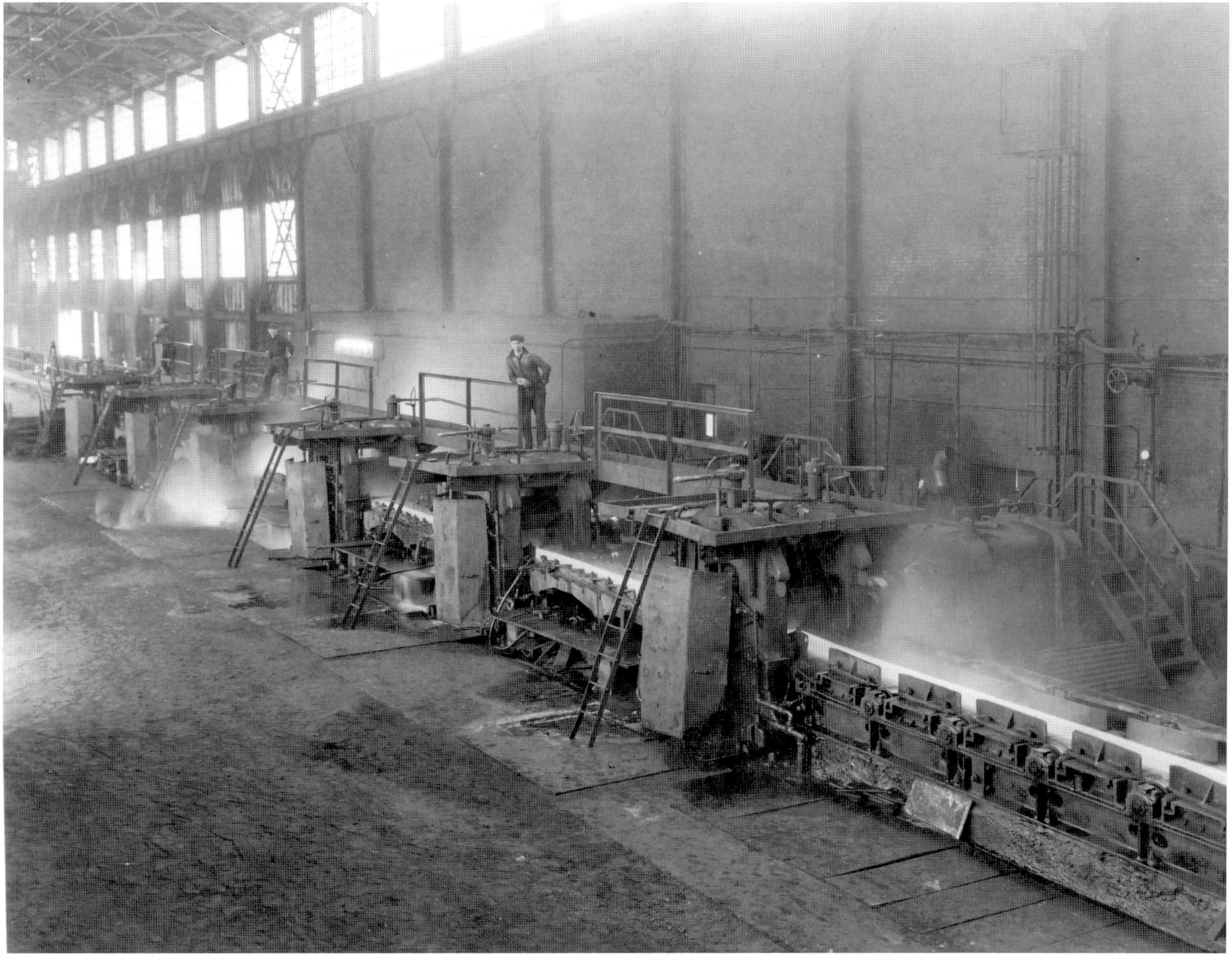

This photo, used as a cover for an issue of *The Otis Sheet,* shows workers standing above the rolls that squeeze the bloom into a long, thin sheet of steel.

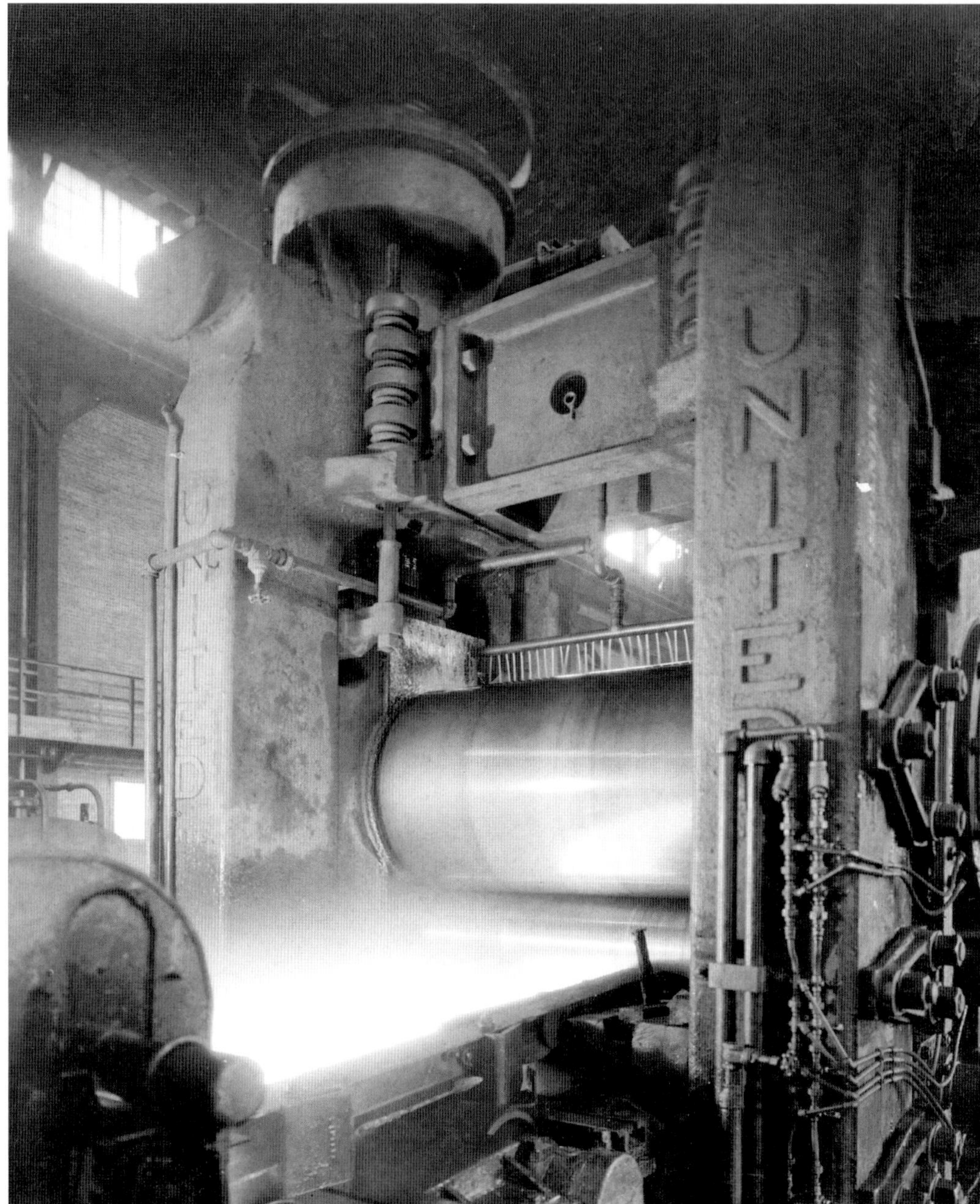

A June 1940 cover for *The Otis Sheet*. This close-up of the roll stand shows the rolls squeezing the steel at high speeds and water sprays cooling it.

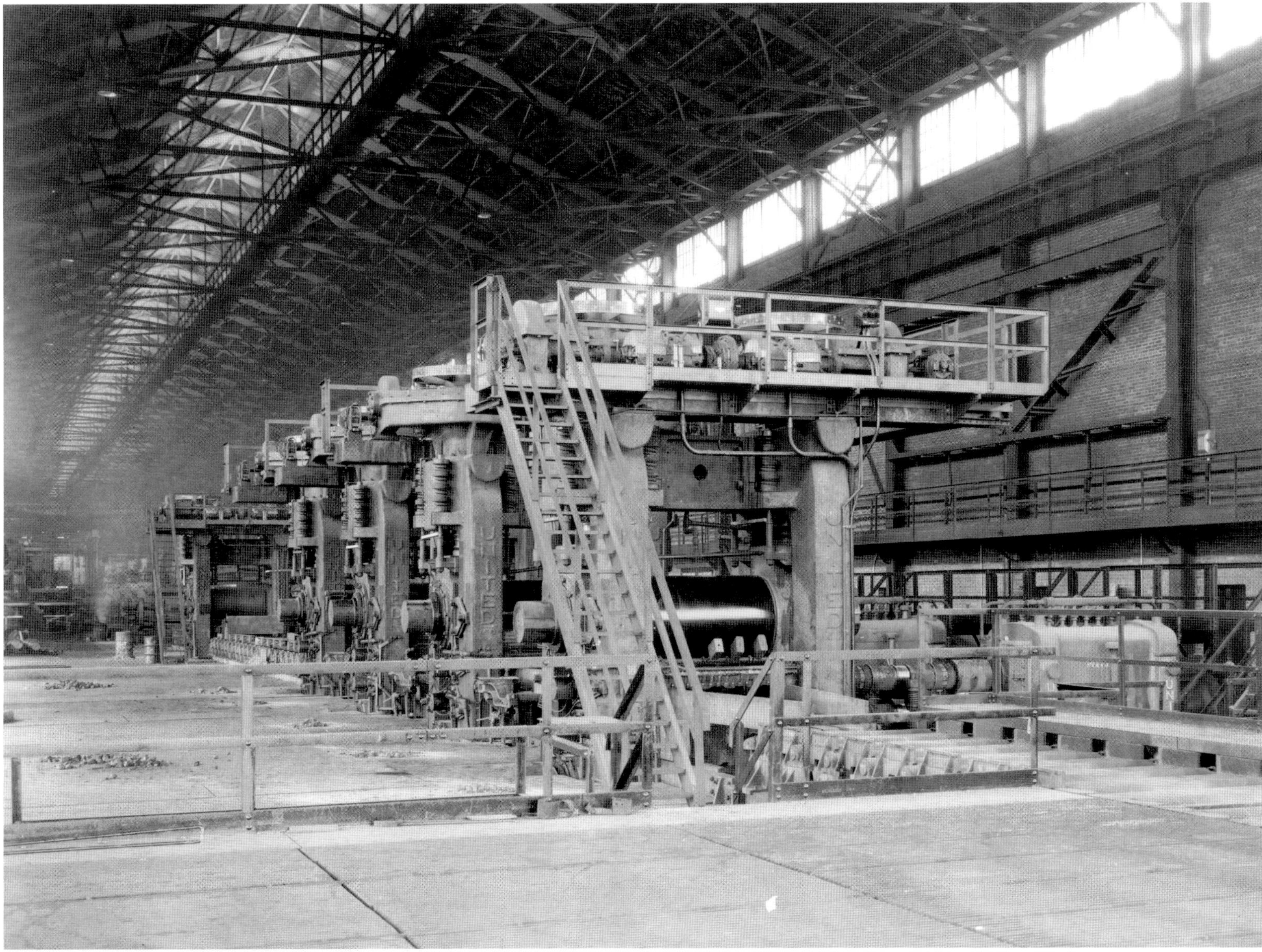

This picture of the hot mill line gives an idea of the scale of the rolling mill buildings, which stretched to incredible lengths and left an indelible impression on observers. At the start of the line, the reheated steel bloom was rammed into the first set of rolls with a titanic crash that often shook the ground in that area.

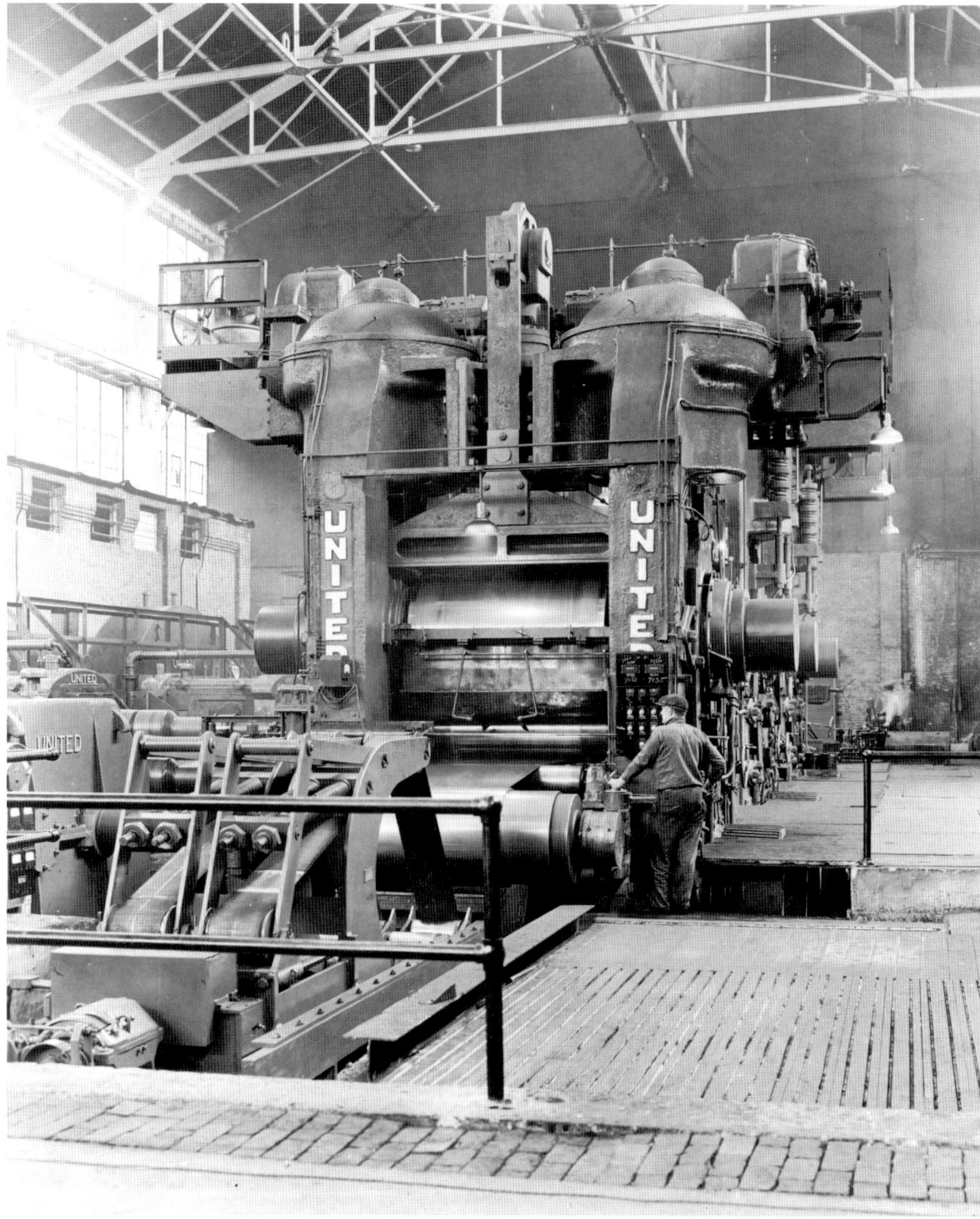

After the steel sheet made its way through the hot mill, it was usually wound into a coil. Coils were often loaded into another rolling mill, a cold mill or finishing mill, which rerolled the steel without reheating it and brought it to the desired thickness.

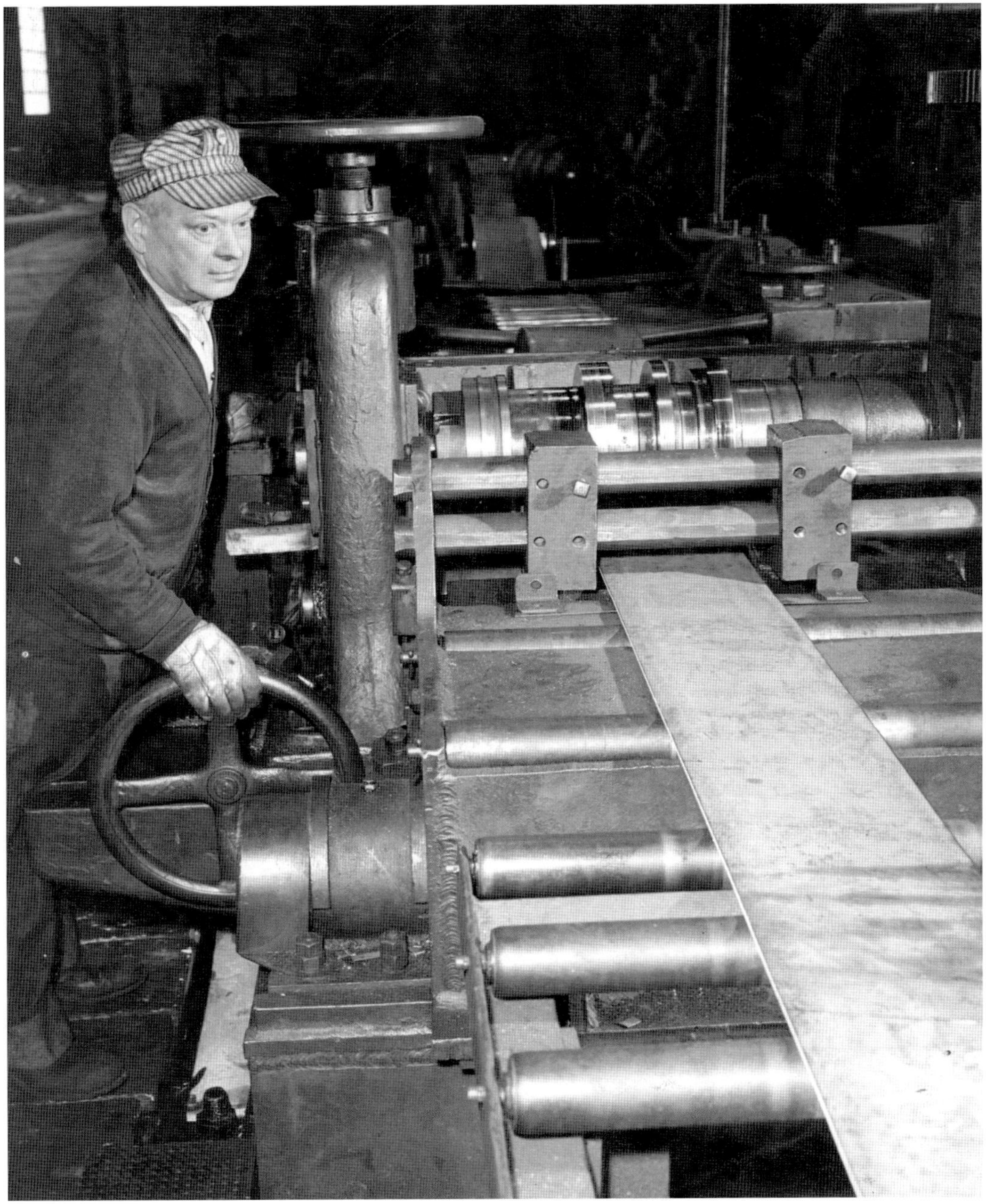

This photo, by well-known Cleveland photographer Herbert Rebman, shows Joe Zelinski, a slitter operator in the 30-inch cold mill department, keeping an eye on the sheet steel, which has been slit to an 8-inch width.

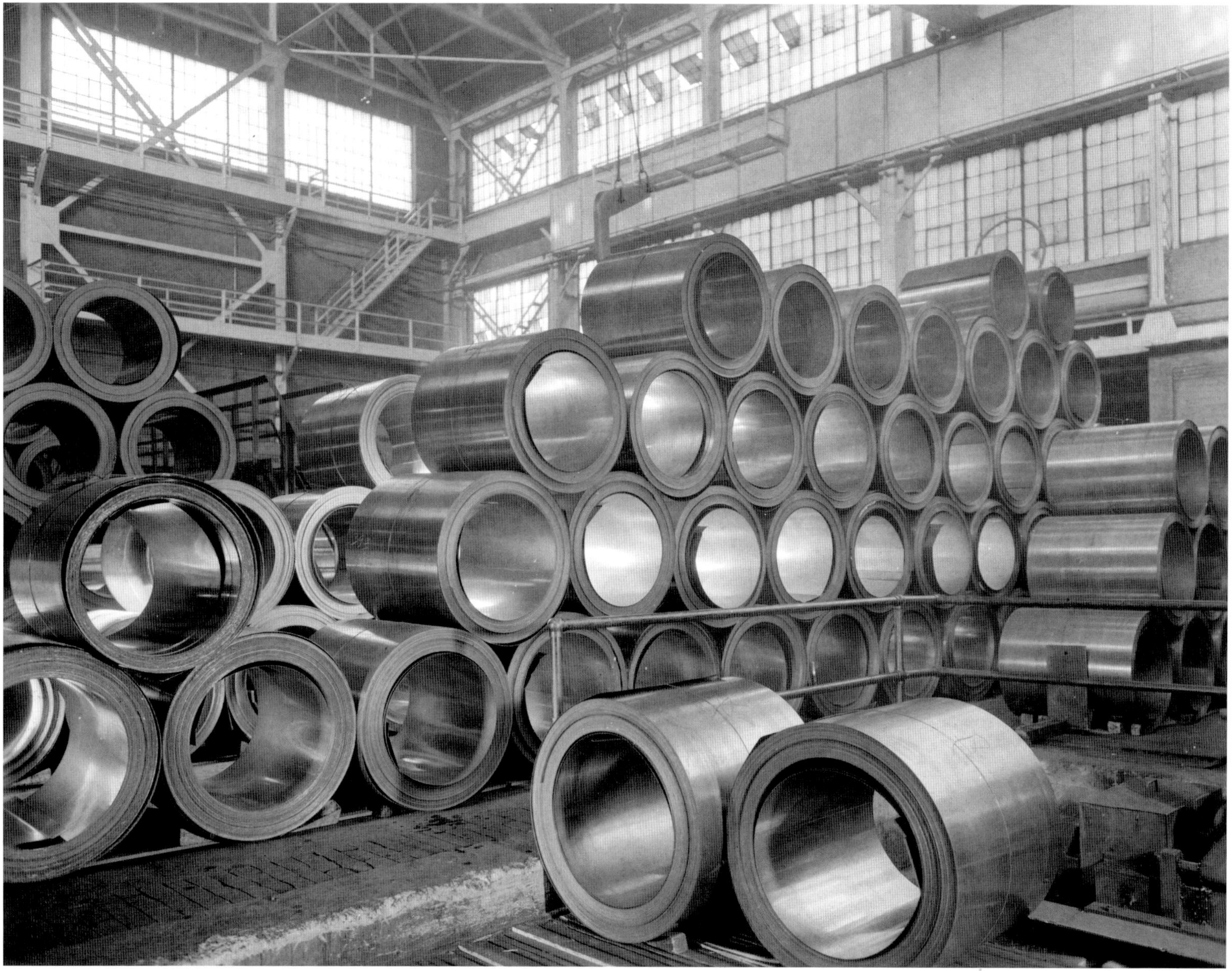

In most cases the steel from the Otis Riverside plant was wound up like these coils from the 72-inch mill. By the 1930s and for decades to come, the auto industry was the major customer of this sheet steel.

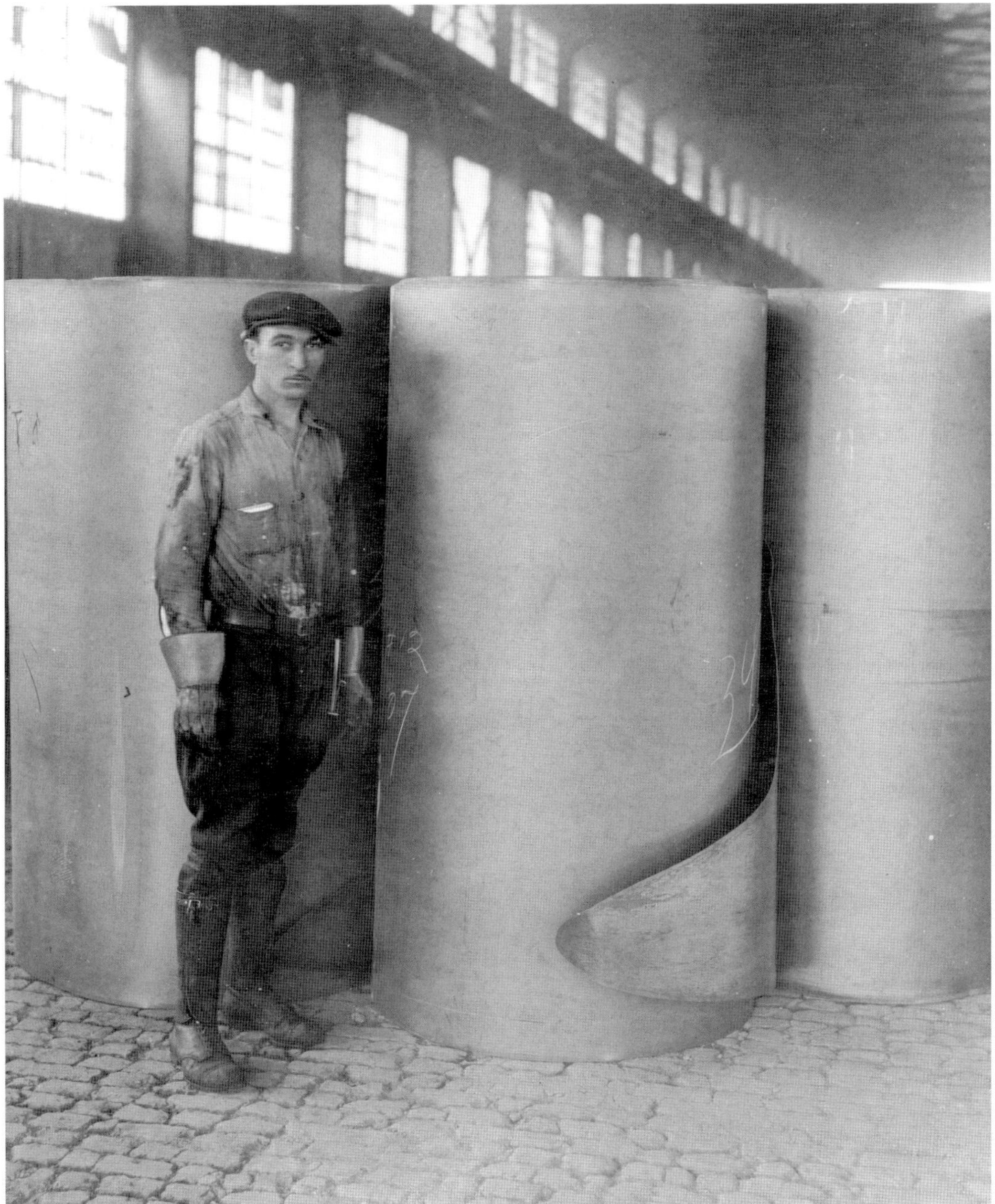

A worker at the 72-inch mill stands next to the finished product.

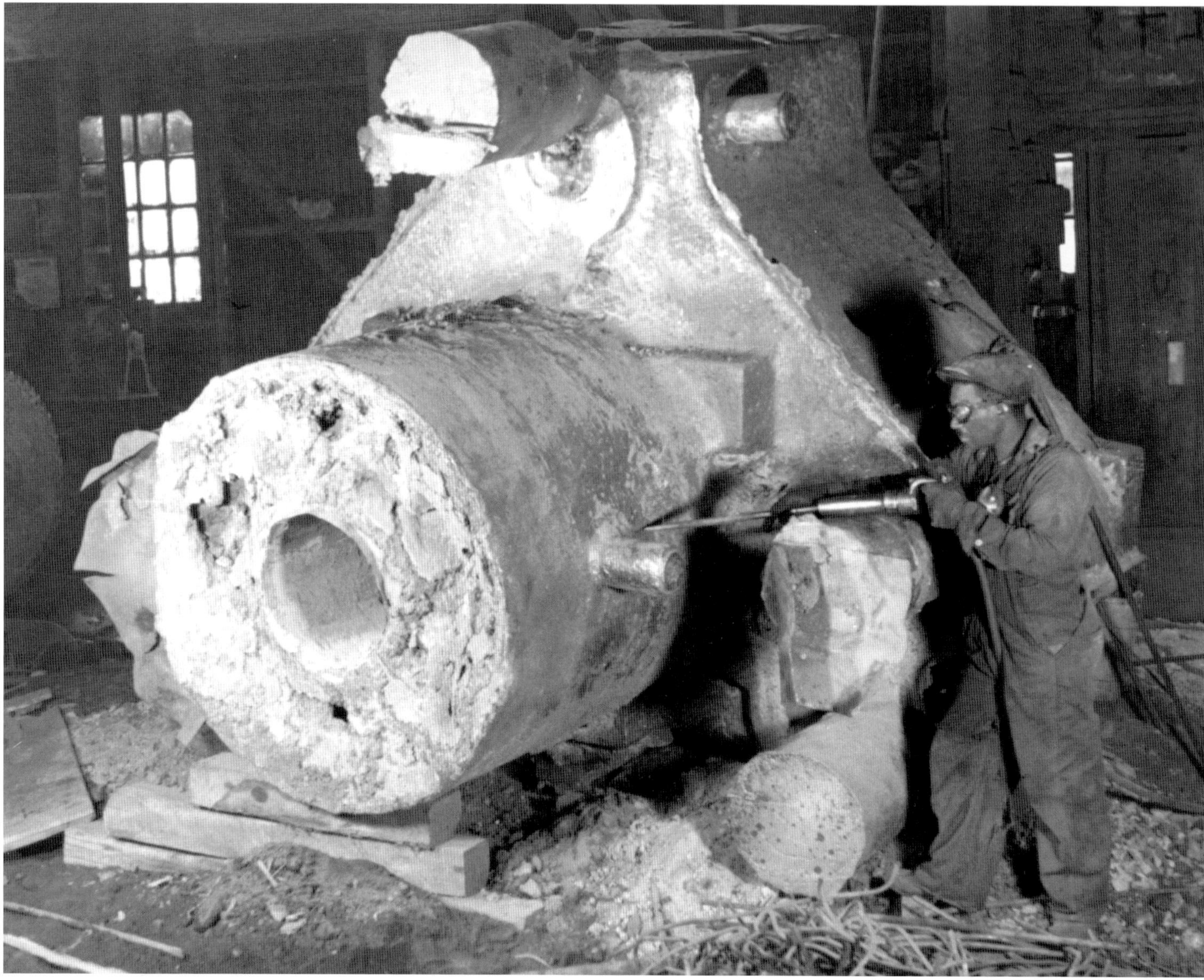

The Otis Lakeside plant, the first mill designed solely to produce open hearth steel in America, focused on large-scale castings, while the larger and newer Riverside plant made sheet steel. This photo shows an Otis worker chipping and cleaning an enormous casting fresh from the mold.

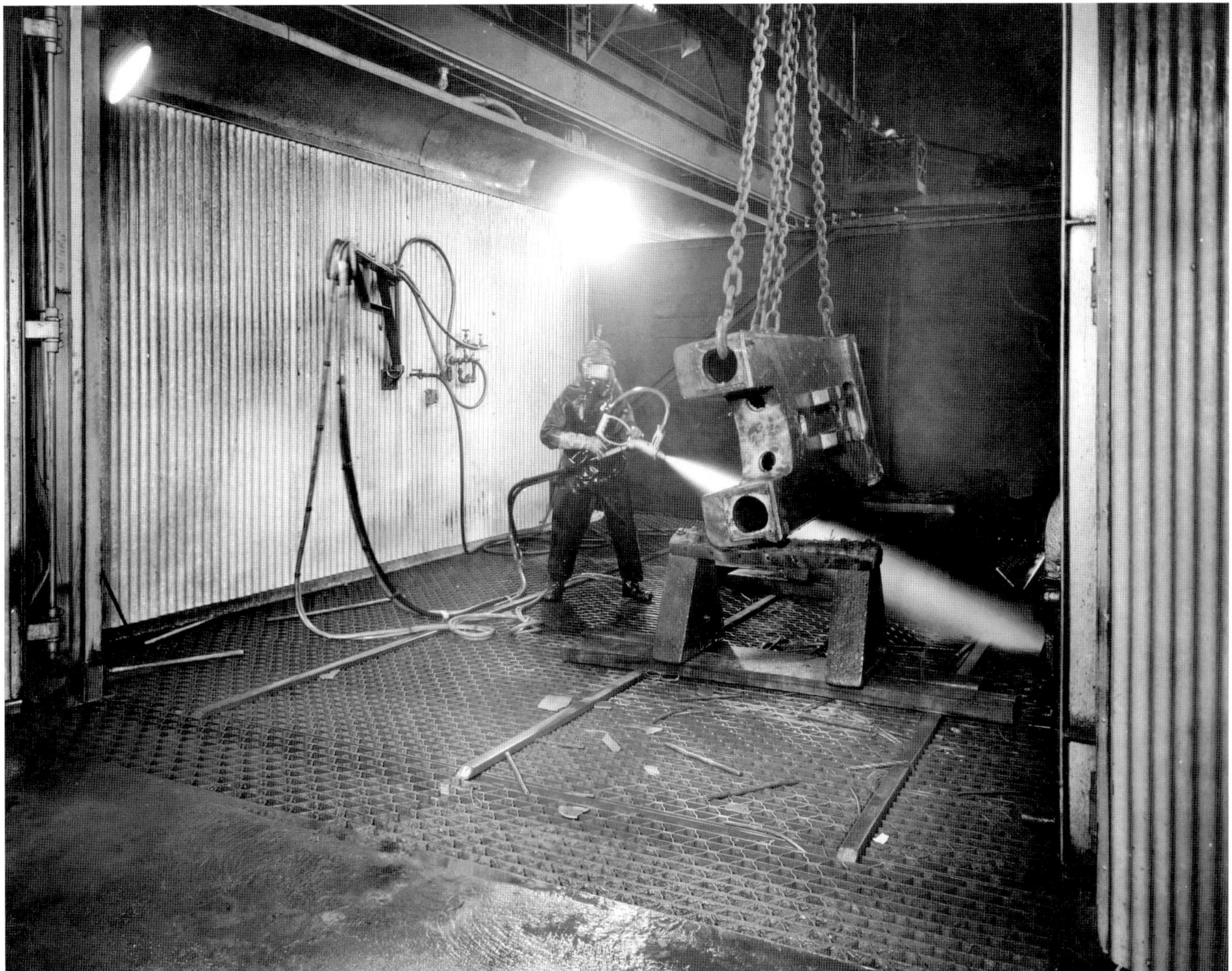

In one of the more unpleasant jobs at the mill, a rubber-suited worker uses high-pressure liquid to blast surface chips off a casting. On a hot summer day the heat inside that rubber suit must have been nearly unbearable.

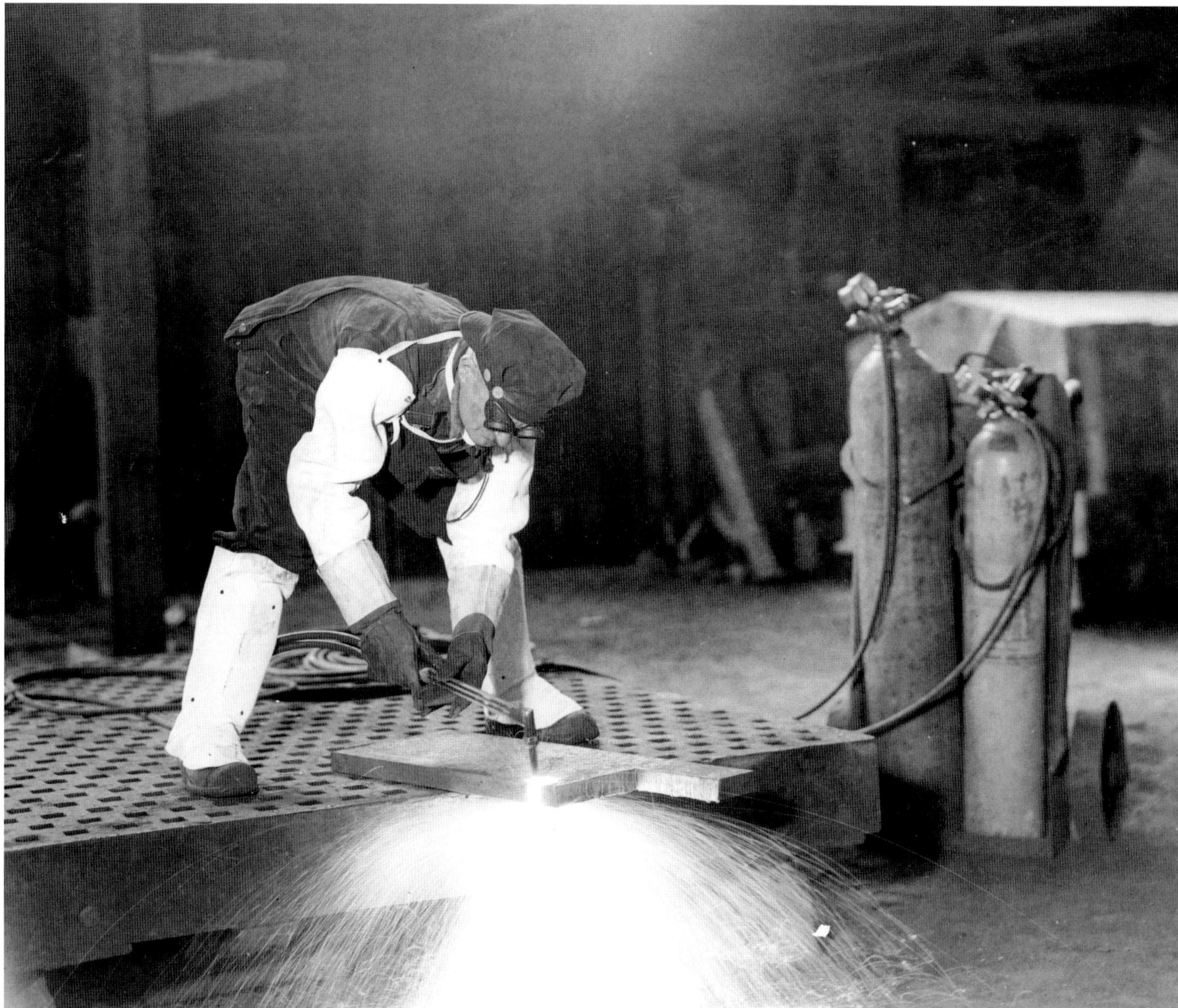

An Otis worker uses a blowtorch to cut a thick piece of steel. While few photos from this era show any safety equipment, this worker appears relatively protected from hazards. Safety helmets were virtually unknown at this time.

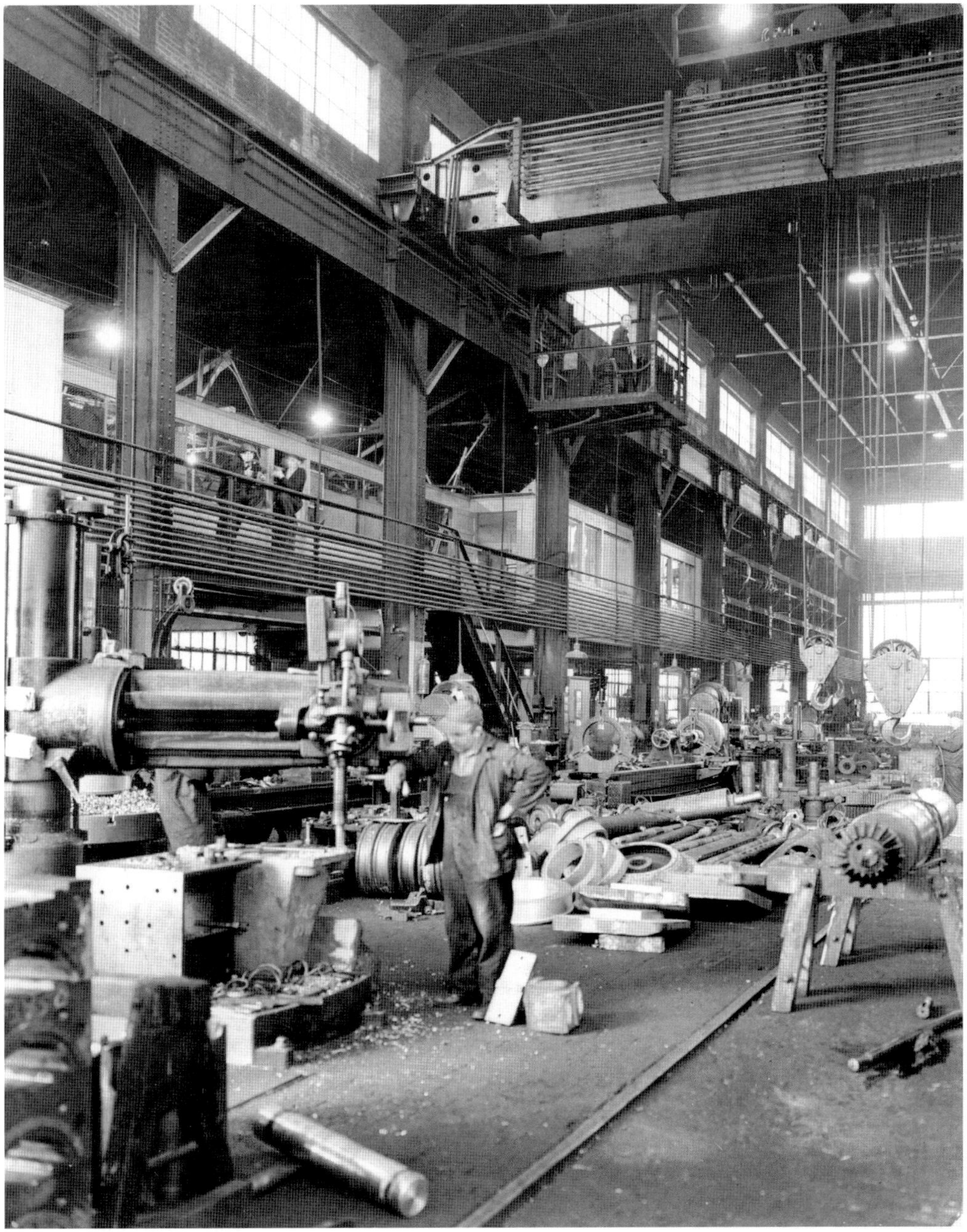

Big machinery requires big tools for repairs, as the giant lathes in the Otis Riverside maintenance shop demonstrate. Photo by Krohn Studios, Cleveland.

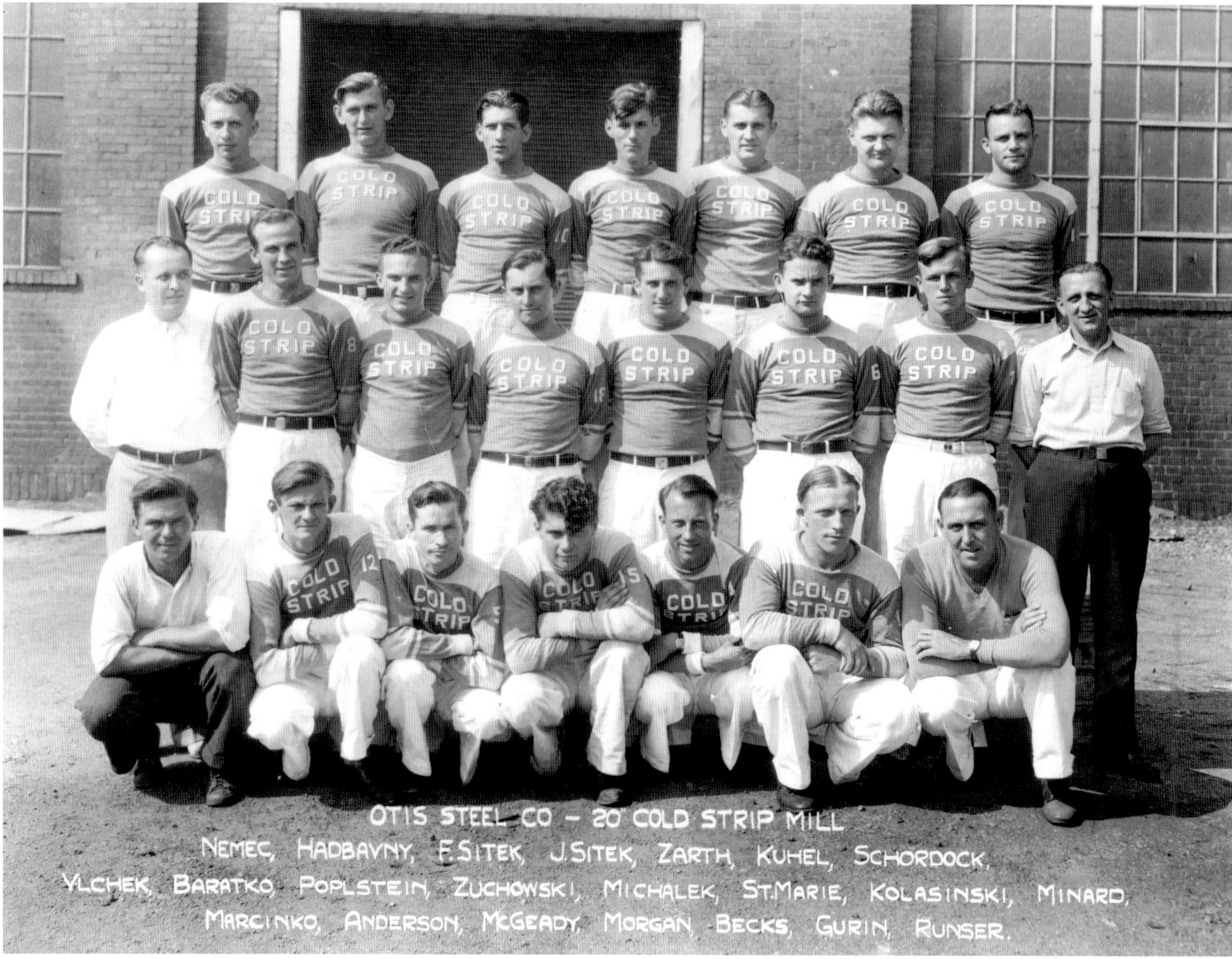

Otis sponsored a variety of athletic leagues that included baseball, football, golf, and bowling teams. This photo shows a team from the 20-inch cold strip mill. The sport is unknown, but the names of the workers show the ethnic diversity among the Otis workforce.

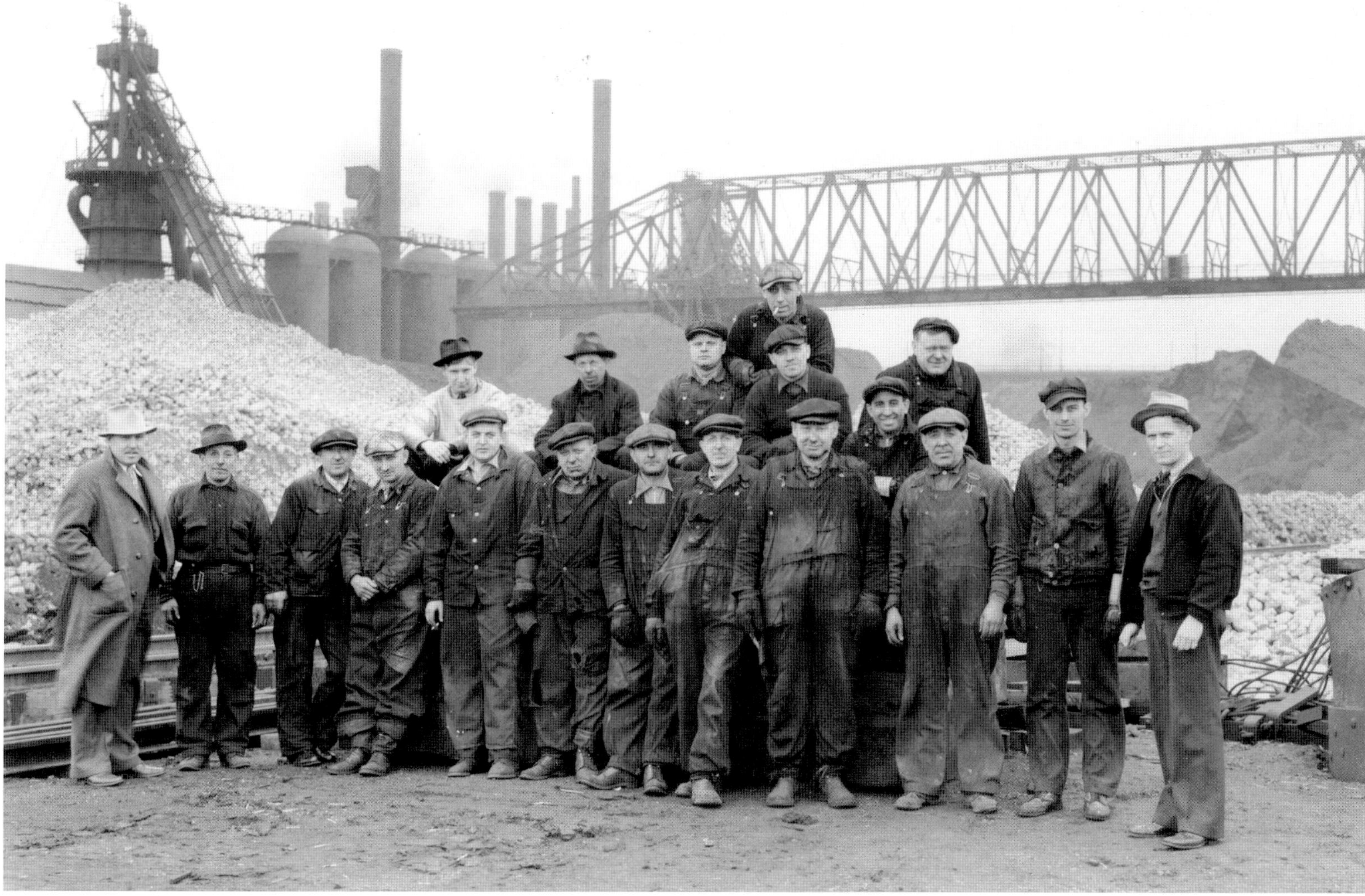

This photo, by Krohn Studios of Cleveland, shows the ore dock crew in 1936. According to the caption on the back of the photo, the workers are: *Front row, left to right:* Harry Sheets (dock superintendent), Stanley Kuchinski (labor foreman), Charlie Senkbiel, Ed Goers, Fillipp Siwik, Joe Naroleski, Tom Jennings, Alexander Abraham, Stanley Novakowski, Steve Kurowski, Anthony Novak, and John Marsinick (dock clerk). *Middle row:* Ed Moran, Mike Lasch, Walter Miller, Carl Hills, Dan Ruscitti, and Ed Vosmik. *Back row:* Dan Ricci.

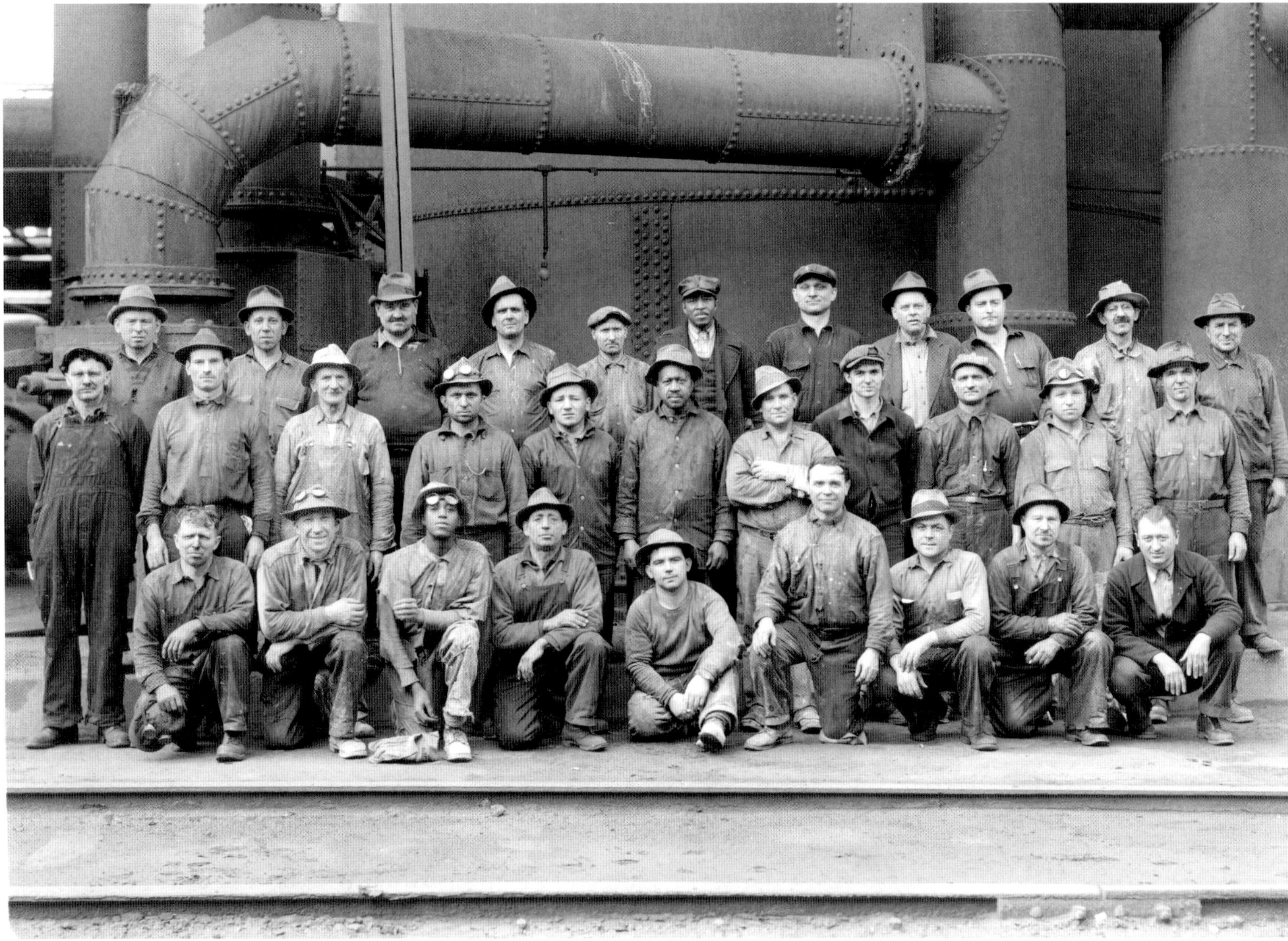

This photograph of the Otis blast furnace crew appeared in the May 1936 issue of *The Otis Sheet,* honoring the crew for operating 426 days without an accident. According to the caption on the back of the photo, the crew is: *Front row, left to right:* Bonazza, Roszkowski, Young, Radzierewski, Wrozak, Kilowski, Marzulli, Neczypor, and DiAngelo. *Middle row:* Unknown, Korostynski, Gonet, Pendalio, Menar, Saunders, Bohonek, M. Blonda, Gebus, Sisko, and F. Blonda. *Back row:* Adamwoski, Gorecki, Nativia, Bilski, Klocek, Evans, Wojton, Swanson, Brown, Muchko, and Kubik.

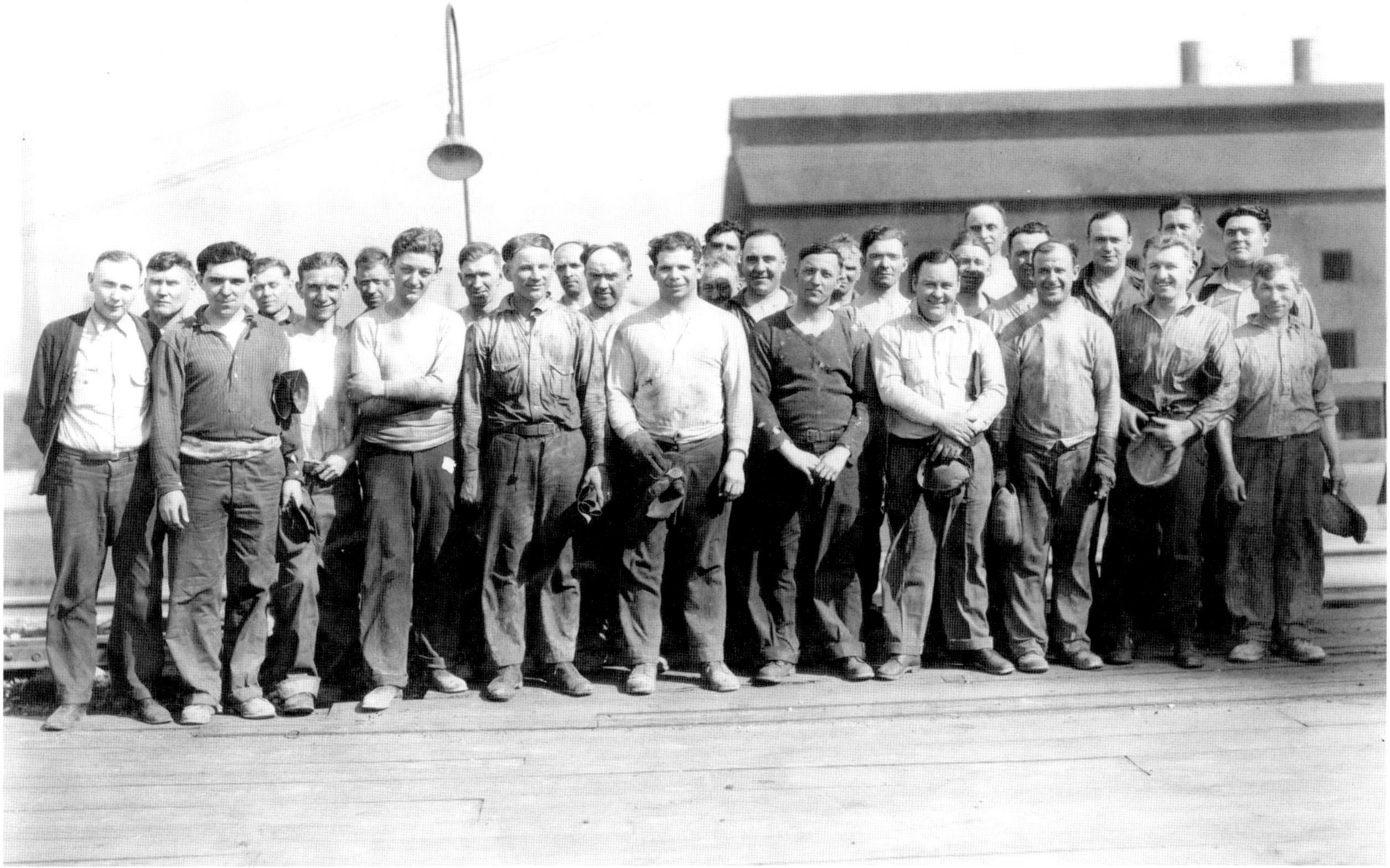

The crew from the open hearth furnaces, ca. 1936. The LTV Steel Collection does contain a nearly complete run of *The Otis Sheet,* but most of the mid-1930s' issues are waterlogged and require conservation work, so it is unknown if this photo appeared in one of those issues.

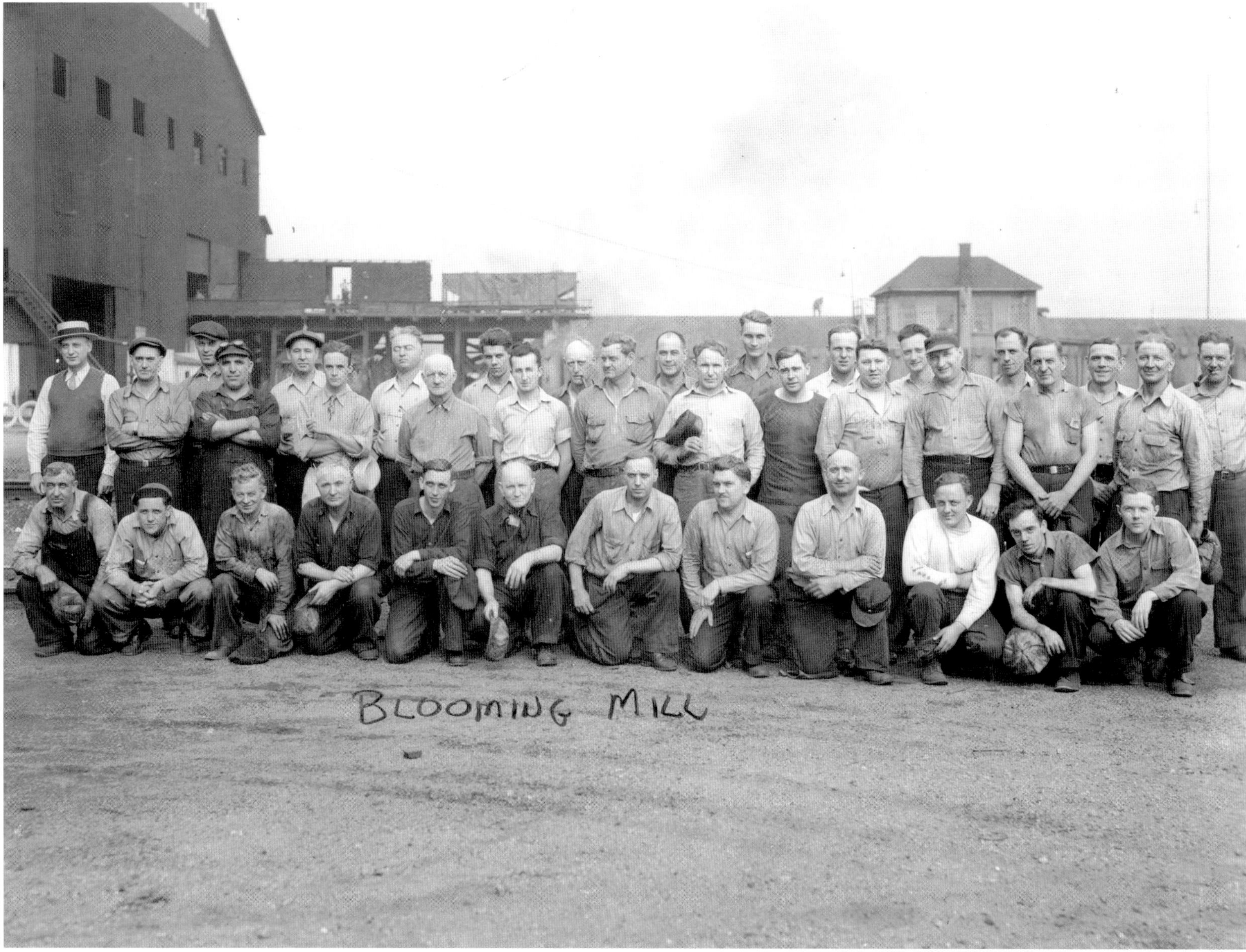

The blooming mill crew, ca. 1936. The caption on the back notes that this is "just a few of the boys who helped the BM safety record."

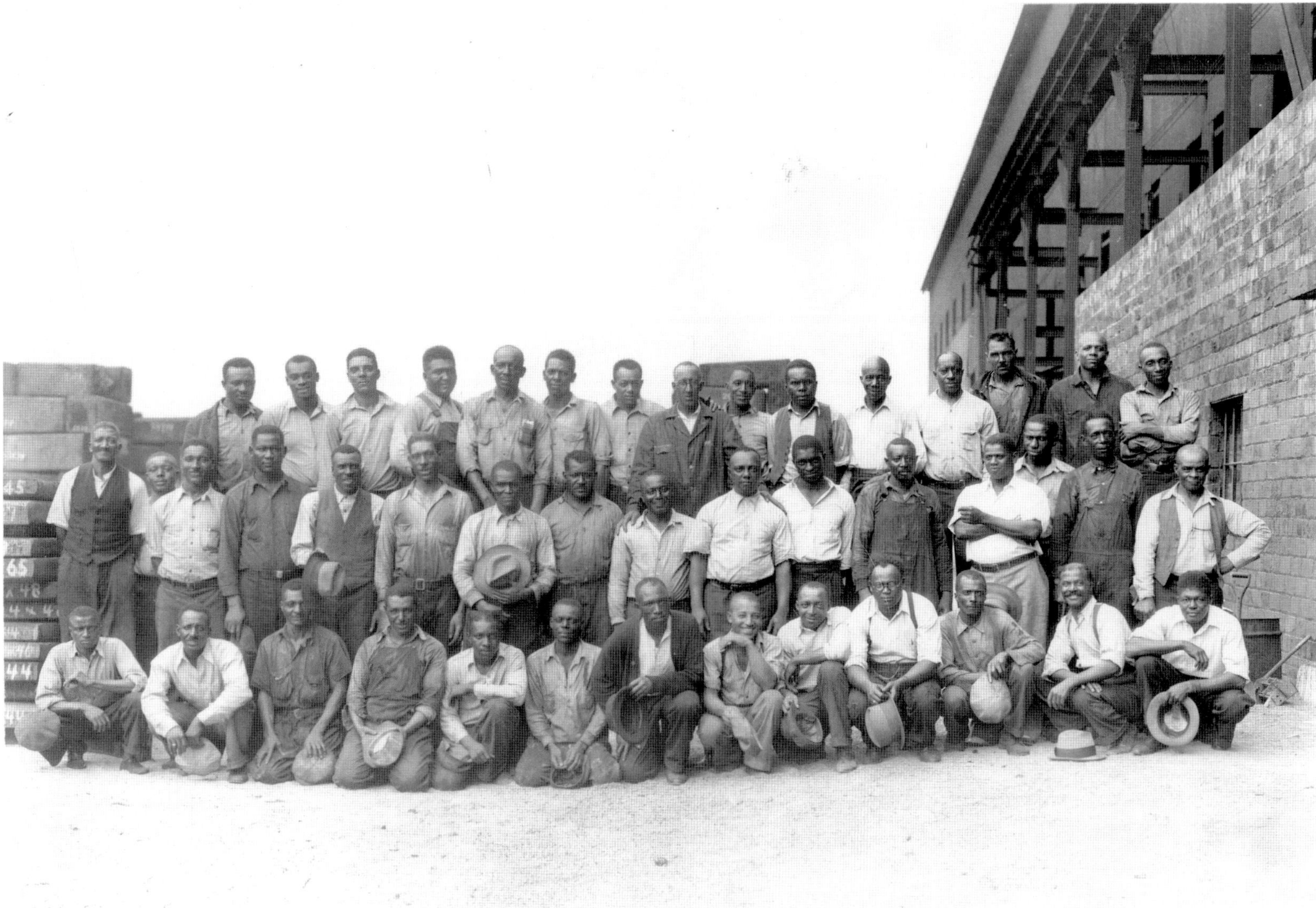

The crew of the blooming mill slab yard is entirely African American, which suggests that their work was particularly unpleasant or dangerous. Photo ca. 1936.

CHAPTER FOUR
REPUBLIC STEEL

Republic was born in depression and reared in adversity. It has never had time or opportunity to get stuffy or stodgy; it has no hardening of the arteries. You sense its individuality whenever you come within its orbit . . . Republic Steel is important to you every day. Your car is made in part of Republic Steel products . . . it is almost certain that you use every day several things made from Republic products. You have an electric refrigerator? The cabinet and part of the motor are likely to be Republic's steel. Republic, like as not, actually manufactured some of the parts of your vacuum cleaner . . . your steel desk, or if not, maybe the filing cabinet or the water cooler cabinet in your office. Your house is piped, not improbably, with Republic's Toncan Iron and your office building has a score of Republic products in its structure. If, by some rare chance, none of these things touch your daily life, there are a thousand other Republic products that may.

—Marc Rose, "Seven Years of Republic Steel." Cleveland: Republic Steel Corporation, 1937

When journalist Marc Rose wrote his celebratory history of Republic Steel in 1937, it looked like the worst was behind the company, and the future looked exceedingly bright. The company had been born at the onset of the Great Depression, yet was succeeding when few observers gave it a chance. However, in less than fifty years, Republic Steel, the third-largest steel company in America, would disappear in a merger with LTV Steel, and ultimately LTV would fail at the dawn of the twenty-first century. And when LTV failed, it did so precisely because it had gotten stuffy and stodgy and had "hardening of the arteries." But Rose was not a seer and could not anticipate what lay ahead. From his position in 1937, he could only look back, and he saw a corporation whose past reached into the early days of the iron and steel industry in America, giving it a foundation for the rosy future he predicted. The new

Republic Steel Corporation dealt with the crushing blows of the Great Depression, weathering them all while growing as a company, and by 1937 it appeared poised for even greater things. With the hindsight of history, it is easy to say that Republic did go on to great things during World War II and beyond, but by the late 1970s, the company had become a poster child for the ailing steel industry and ultimately disappeared in a consolidation, much like the one that had spawned its namesake predecessor. But Rose couldn't have seen that, and nothing in Republic's interesting past could have given him much sign of what the future would really hold, that a company born of consolidation might eventually die of consolidation.

The Republic Iron and Steel Company was incorporated in New Jersey in 1899. At the time, the iron and steel industry was undergoing a massive wave of consolidation, as smaller companies were swept up in large trusts, such as American Sheet & Tin Plate Company, Federal Steel Company, American Steel Hoop Company, and the largest and mightiest, United States Steel Corporation. While these consolidations were going on, a number of smaller iron and steel companies that were not part of these business moves consolidated into a company called the Republic Iron and Steel Company. Forty companies made up this new corporation, though some observers at the time noted that of all the new steel trusts, this was by far the weakest, as it was made up of a number of marginal companies, many of which were operating in receivership and others of which had wholly obsolete facilities.

The new corporation addressed this weakness by restructuring itself so that the newest and most advanced mills became the backbone of the company, while obsolete mills were dismantled or otherwise closed. The company pumped money into the existing plants to renovate and update their facilities. It also sought larger coal and ore reserves. Profits were initially elusive, but Republic Iron and Steel persisted, pouring what money it could into expanding and updating its facilities and adding blast furnaces and new larger Bessemer converters. Sales increased slowly and profits began to grow, as the company's new and renovated facilities increased production levels. The outbreak of World War I in 1914 threw the company into disarray, and 1914 proved to be one of the worst years in the young company's history. By 1916, however, profits were soaring and production was at a peak because of wartime orders. After the war the company began a slow expansion, acquiring a tin plate company in Niles, Ohio, and buying up more iron and coal reserves. During the 1920s Republic Iron and Steel continued updating its facilities and making money at a steady pace. By the late 1920s, however, a new pattern emerged: the company began rapidly expanding by acquiring other companies. This was a radical departure from the slower growth pattern that the company exhibited in its first decades of life, and was due to Cyrus Eaton, the company's largest shareholder by 1927.

Often overlooked today, Eaton caused much excitement with his bold financial moves in the late 1920s. Born in Nova Scotia, Eaton came to Cleveland in 1907 to visit his uncle, a noted local minister. Eaton's uncle was an acquaintance of John D. Rockefeller, and the oil titan soon met the young Canadian. Much impressed with the young Eaton, Rockefeller soon hired him, and Eaton became his protégé. It wasn't long, however, before Eaton, relying on his sharp business acumen, struck out on his own. After setting up the Canada Gas and Electric Corporation in 1912, Eaton returned to Cleveland and became a banker. He later organized an investment trust called Continental Shares, which he used to purchase interests in industries he felt needed consolidation to advance. Soon he entered the steel industry, gaining stock control of various companies and combining them. With Republic Iron and Steel, he reached out and acquired nearby companies such as Trumbull Steel Company, Trumbull Cliffs Furnace Company, Steel and Tubes, Inc., and Union Drawn Steel Company. The moves were smart, as these companies could further process and use the steel from Republic's steel-producing facilities. Furthermore the acquisitions were hardly marginal; Trumbull Steel had a modern and efficient plant that produced strip steel, and Steel and Tubes was one of the top producers of electric-welded pipe, giving Republic a significant presence in a growing market. With its diverse product line, Union Drawn would be a ready consumer of Republic steel.

Those moves alone would ensure Republic Iron and Steel a fairly solid future, but Eaton was not through. His goal was bolder. He wanted to create a Midwest steel empire that would rival or even surpass the mighty United States Steel Corporation. He turned to several more companies, including the newly formed Central Alloy Steel Corporation in Canton, which was made up of the former United Alloy Corporation of Canton, Central Steel Company of Massillon, Berger Manufacturing Company of Canton, Interstate Iron and Steel of Chicago, the Donner Steel Company of Buffalo, and the Bourne-Fuller Company of Cleveland.

The merger was announced on December 18, 1929, and caused a storm of headlines in the newspapers in Northeast Ohio, where most of these companies were located. It was also announced that Republic Iron and Steel would be renamed the Republic Steel Corporation. The *Cleveland Plain Dealer* quickly pointed out that while this merger created the third largest steel company in the country, it was only a first step, since Youngstown Sheet and Tube and possibly Inland Steel would join the merger. Other rumors suggested Jones and Laughlin would eventually join, and Republic would be the largest steel company in America. Joining Eaton in this endeavor was Elmer T. McCleary as president of Republic and Tom Girdler as chairman of the board. McCleary had been with Youngstown Sheet and Tube for twenty-two years and had risen from a chemist and metallurgist to vice president. He had a strong reputation throughout Youngstown for being quiet and unassuming but direct. Even though he was their vice president, many longtime workers at Youngstown Sheet and Tube called him "Mac" to his face, and he still spent much of his free time working with the Youngstown YMCA.

The new chairman, Tom Girdler, was cut from a different cloth. Gruff and profane, he had been president of Jones and Laughlin before assuming his role at Republic. Girdler was an engineer by trade who had worked his way up the steel industry. Employed by various companies until 1914, Girdler joined Jones and Laughlin as an assistant general superintendent of the Aliquippa Works in Pennsylvania, where he worked his way to the top. He ran Aliquippa's plant with an iron fist, and the town became known as "Little Siberia" for the control the company exerted, even in community affairs. But Girdler was also an organizational genius and widely considered one of the best leaders in the steel industry. He had corresponded and met with Cyrus Eaton before the consolidation. Eaton knew of Girdler's reputation at Jones and Laughlin and wanted him for the company he had planned to form. By many accounts, Girdler was close to jumping over to Eaton's side, when the president of Jones and Laughlin suddenly died, and the families controlling the company offered him the presidency. He accepted, but in less than a year was gone. According to Girdler's account, he quit. According to Eaton's account, forty years after the event, Jones and Laughlin fired Girdler, and he literally came crying to Eaton. Regardless of what really happened, the result was that Tom Girdler became Republic's chairman of the board and was charged with making the acquisitions work

Girdler's work was not easy. His first challenge was replacing Elmer McCleary, who died suddenly in April 1930. The stock market crash and the onset of the Great Depression had radically changed the business climate in the United States, and Cyrus Eaton, the man first behind the consolidation, had lost his fortune in the stock market and dropped out of Republic Steel's leadership. The new company was facing its most difficult test right from the start. Taking over as president, Girdler assembled a team of hard-boiled executives, all of whom had extensive experience in the steel industry and were cut from the same cloth as Girdler. Many were veterans of Jones and Laughlin, so Girdler already trusted them. Because money was tight, some operations were curtailed and employees were let go. Executives spent time working the phones and back doors of banks, soliciting loans and financial support to keep the young company from going under. It didn't help that many of the banks were going under too.

But Republic Steel was blessed with a few things to keep it afloat. One was its variety of products, such as the alloy steels from its Central Alloy Division, and its lighter finished steel products for the growing automotive industry. The company's state-of-the-art electric-weld pipe mill was also a moneymaker. In addition Republic had excess inventory from the merger, which could carry it through lean times. However, as Republic worked through its growing pains and tried to survive the Depression, it still faced another problem: the bulk of its plants were located in Ohio's Mahoning Valley, one of the country's largest steel-producing areas, but one hampered by high transportation costs. Unlike plants in Chicago, Cleveland, Buffalo, and Pittsburgh, the Mahoning Valley plants were essentially landlocked and therefore relied on railroads for transportation. Water transportation was far cheaper than rail, so companies using water transportation for raw materials and finished products had a distinct advantage. Tom Girdler went looking for more mills to add to Republic. Two on Lake Erie caught his eye: Otis Steel and Corrigan, McKinney in Cleveland. Otis was dismissed for its older equipment and high price, but Corrigan, McKinney was a different matter. It had long been an efficient and cost-effective mill, and Girdler wanted it for Republic. The federal government was opposed to the acquisition. Republic argued that it needed more capacity for semifinished steel and that further consolidation would stabilize the market and be better for the market rather than adding capacity through new construction. The deal went through, giving Republic a giant steelworks on Lake Erie, dock facilities (which saved the then-dockless Republic a good deal of money), a Corrigan, McKinney–owned sheet steel mill near Detroit, and dramatically increased coal and ore holdings. Girdler also boosted Republic's product mix by adding Youngstown's Truscon Steel, the country's largest manufacturer of steel building products, with warehouses and sales offices around the country.

In 1935 Girdler and his staff relocated to Cleveland, site of Republic's newest acquisition, and established headquarters there that remained for the rest of the Republic's existence. When *Fortune* magazine examined Republic in 1935 and commented favorably on the acquisition of Corrigan, McKinney, it envisioned a time when Republic would shutter all its Mahoning Valley plants and instead concentrate production in Cleveland, because of transportation costs. The magazine's predictions proved eerily accurate. In the end, all of the Republic plants in the Mahoning Valley area were either sold off or closed after the LTV merger and bankruptcy; the former Corrigan, McKinney plant in Cleveland was one of the sole survivors.

The new moves were helping improve Republic's bottom line. In 1936 the corporation had a net profit of $9.5 million, but in 1937 a new challenge set the company back, as the resurgent labor movement targeted Republic for a major strike. Republic was one of the "Little Steel" companies, so named because United States Steel was considered solely "Big Steel," and none of the other companies, massive as they were, could compare. "The Corporation," as U.S. Steel was often called, was not a target of these strikes, because it had already made a deal with John L. Lewis, head of the Committee for Industrial Organizations (CIO). U.S. Steel was beginning to pull out of its Depression slump and did not want labor strife to put that at risk. Little Steel was a different matter, as many of the Little Steel company heads were virulently antiunion. Republic's Girdler, Bethlehem Steel's Eugene Grace, and National Steel's Ernest Weir were all steadfast in their opposition to labor unions. And so the CIO sent union organizers into the steel towns to try to finally get the steel industry unionized.

The fight was bitter and violent, and Republic was part of some of the bloodiest battles of the entire strike, including one of the most violent uprisings on Memorial Day 1937. Striking workers and their families marched on the Republic Steel plant in Chicago that day, and local police guarding the plant charged the strikers, throwing tear gas grenades and beating workers and their wives with billy clubs. As the strikers and their families turned and ran, policemen opened fire. Ten people were killed and dozens wounded. Immediately the company and the media declared the march was made up of Communists. However, a later Congressional investigation concluded that the marchers were not Communists, and newsreel footage shot at the event clearly showed policemen firing at the backs of fleeing workers and beating female family members. Eventually during World War II, Republic quietly settled with survivors of the event and family members of those killed, paying out settlements ranging from a few hundred dollars to a few thousand.

The violent strikes in Chicago and other Republic cities demoralized workers and had to be ended. Unable to organize the Little Steel mills, the CIO persisted because it knew it had the advantage: the federal government had become far more supportive of labor than in the past. Eventually Republic and the other Little Steel companies were unionized, due in large part to the government's insistence. The other factor was the war. Girdler and the other antiunion leaders didn't want to miss the huge profits that war production would bring their companies.

By 1937 Republic was finally turning the corner financially and able to embark on an extensive program to modernize and expand its plants. In Cleveland, near the newly acquired Corrigan, McKinney Works, the company built the world's largest hot strip mill (98 inches), which rolled wide strip steel for the automotive industry. Building this mill entailed straightening and slightly moving a long stretch of the Cuyahoga River. In Niles, Republic added a new mill for cold-rolling and finishing tinplate, while Chicago received a new wire mill. The company also expanded the electric-weld pipe works in Youngstown, the silicon strip steel production in Warren, and stainless steel strip production in Massillon. To expand geographically, Republic purchased the Gulf States Steel Company in Alabama.

This expansion and modernization served Republic well when the United States entered World War II. The company had good facilities to make the quantity and quality of steel needed for the war effort, and like its predecessor, Republic Iron and Steel, the new Republic Steel sent a good portion of its workforce, more than twenty thousand workers, into uniform to fight in the war. To make up for the loss in workers, retirees returned to the mills and the company hired large numbers of women for the first time. Republic even participated in a War Department movie, *Women of Steel,* which described the many jobs in the mill that women workers could undertake. However, the film made a point of emphasizing that all those women would return to the kitchen as soon as the war was over. Those women, doing some of the dirtiest and heaviest work, helped their male counterparts go the extra mile for the war effort. In one of the Canton facilities, women composed 90 percent of the workforce.

Expansion was the story of postwar Republic Steel. Tom Girdler remained chairman of the board until 1955, but in 1945 his protégé, Charles M. White, took over operations. Under White the company embarked on expanding and modernizing many of its facilities. A key element to Republic's plans in the 1950s was to improve the company's iron ore sources. Republic acquired ownership of a company that had an iron ore concession in Liberia and aggressively developed iron ore reserves in northern Canada. One of the company's best moves was to join with competitor Armco Steel in acquiring the Reserve Mining Company of Minnesota. The two companies built facilities to process taconite, a low-grade ore that was formerly viewed as waste material. As sources of high-grade iron ore declined, taconite appeared to be a viable alternative, as long as a way could be found to extract the iron ore from the taconite rock. A process was eventually developed in which taconite rocks were crushed into marble-sized pieces, mixed with water, and ground into a powder by rotating mills. This process allowed mines formerly viewed as depleted to become productive again, and using taconite became commonplace in the industry.

Republic continued to improve its technology and processes in the 1960s under the direction of Thomas Patton. The open hearth process was being superseded by the new basic oxygen process, which greatly reduced the time needed to make a heat of steel. The company's plants in Warren, Gadsden, and Cleveland were the first to get this technology in 1965 and 1966. By the end of the decade, 90 percent of Republic's facilities had been upgraded or newly constructed. While sales grew, costs also grew, and profits proved more elusive. Imported steel became increasingly competitive. In 1971 Republic had one of its worst years ever, with earnings at its lowest since the 1930s. Profits rebounded from 1972 to 1974. However, imports also increased, as did government pollution regulations, which became a major drain on the company's capital investments. The high energy costs of the late 1970s also hit Republic hard. The company compensated by investing in new sources of coal and gas for its facilities and implementing conservation practices.

Republic campaigned in Washington and through the media against steel imports. In 1982, however, all the price and profit fluctuations of the 1970s, the soaring energy costs, the increased labor costs, and the increases in imports finally came to a head. Suffering an economic slowdown, the industry as a whole faced massive financial losses and personnel layoffs. It looked like the hemorrhaging would never end. In 1982 Republic laid off more than 5,000 workers from its mills in Canton, Massillon, Cleveland, and Warren and shut down various departments at a number of mills. In early 1983, with an expected turnaround still not occurring, another 17,000 workers were laid off or put on reduced work weeks, while the salaried workforce was reduced by more than 3,000 employees.

While it appeared that the brutal cost-cutting measures of 1982 might have helped the company weather the storm, in 1983 talk at Republic soon turned to a merger with Jones and Laughlin's parent corporation, LTV. The United States Department of Justice opposed the merger, just as it had when Republic wanted to acquire Corrigan, McKinney in the 1930s. However, both companies negotiated, and an agreement was reached. Republic would divest itself of the Gadsden plant and the Enduro stainless steel line in Massillon. In 1984, with the Department of Justice's approval, Jones and Laughlin (which had merged with Youngstown Sheet and Tube in 1978) merged with Republic Steel, and the new company became known as LTV Steel. The result was the second-largest steelmaker in America, surpassing Bethlehem Steel, and essentially completing Cyrus Eaton's dream of fifty years earlier. The new company was headquartered in Cleveland. Despite the merger, the steel market was not improving; prices continued to plummet, and the giant new company continued losing money. In 1986 the entire LTV Corporation filed for reorganization under the Chapter 11 bankruptcy laws. More and more facilities were sold off or closed, including the old Republic Steel plant in Warren, the Republic Steel bar lines in Canton and Massillon, and other LTV interests, including aerospace, military vehicles, missiles, electronics, sporting goods, and food businesses.

The LTV that emerged from bankruptcy in 1993 was primarily a steel and energy company (though eventually the energy company was sold), rather than a giant conglomerate that owned steel mills and countless other interests, many of which were utterly unrelated. During the bankruptcy, the company ruthlessly shed facilities left and right. Before bankruptcy, LTV had twenty-eight major steel plants. Afterward there were only thirteen major plants, two of which were integrated steelmaking facilities, the former Republic and Jones and Laughlin mills in Cleveland now united as a single facility, and the former Youngstown Sheet and Tube facility in Indiana Harbor, Indiana.

But LTV, born in desperation, found that it could still not escape problems that had plagued its predecessors. The company had efficient plants, but management seemed unsure of how to operate them. The industry was back in a cycle of good years and bad years, but LTV kept spending money on acquisitions, while the big, integrated mills kept losing money. All this time steel imports continued to climb and minimills took away market share. In 2000 the bottom fell out for the last time. Prices for flat-rolled steel, LTV's big product, dropped 35 percent. LTV quickly started cutting jobs and closing facilities, and on December 29, 2000, the company filed for bankruptcy again. Though the company's executives swore they could turn the company around, the company continued losing money and began selling off various operations.

By 2002 there was no turnaround plan. LTV was at an end, and, as part of the bankruptcy process, the company agreed to sell all its assets. A number of suitors appeared on the scene, including U.S. Steel and Bethlehem Steel, but an unlikely bidder, Wilbur Ross, made the best deal. Other bidders wanted pieces of LTV, but Ross wanted the whole company intact. An investor who specialized in turnarounds, Ross won the LTV facilities in March 2002 and quickly set up the International Steel Group to run them. The new company took over the old LTV mills and quickly restarted them, negotiating an innovative and flexible labor contract with the union workers to operate the plants more efficiently. Suddenly the mills that couldn't make money became profitable. And ISG was poised for more, as giant Bethlehem Steel also entered bankruptcy and soon sold its assets to ISG, as did Weirton Steel, a tinplate producer. In just two years of existence, ISG became the biggest steel company in America, surpassing "The Corporation," United States Steel, in one of the biggest waves of consolidation since the formation of U.S. Steel a century earlier. Based in Cleveland, the company perhaps finally fulfilled Cyrus Eaton's vision of a massive midwestern steel empire. But even that lofty perch was short-lived, as Wilbur Ross, the financer who had turned around the remains of a shattered LTV, decided to move on. At the end of 2004, multinational steel firm Mittal Steel purchased ISG and became the largest steelmaking company in the world.

While Republic Steel has been gone for more than twenty years, pieces of the company remain. Several Republic Steel facilities continue to operate under different owners, including the Cleveland mill that is part of Mittal Steel. When LTV merged with Republic Steel, workers bought the bar division of Republic through an employee stock ownership plan and entered business as Republic Engineered Steel. Later, after a Mexican steel producer acquired it, the company was renamed Republic Engineered Products, and still operates former Republic Steel facilities in Canton and Massillon. Another survivor is Republic Storage Systems. Like the bar division, it is a former part of the Berger Manufacturing Company, one of the major acquisitions that created Republic Steel Corporation. After the LTV merger the employees of the storage systems division bought the division in 1986 and began operating it independently. Republic's Warren Works, initially the Trumbull Steel Company that was acquired in 1928 by Republic Iron and Steel, was sold off in 1988 and became WCI Steel. Although it has undergone its own bankruptcy and ownership changes, WCI Steel still survives. Truscon Steel, purchased in 1935 by Republic, also operates independently as Windsor Republic Doors, having been sold off by Republic Steel in 1985.

It seems that the very assets that Cyrus Eaton and Tom Girdler felt were necessary to make their steel giant succeed are still viable producers today. Republic's corporate life was short and tempestuous. Reared in adversity, as Marc Rose wrote in his history of the then seven-year-old company, Republic survived the Great Depression, the labor turmoil of the late 1930s, World War II, and the Cold War. A child of capitalism, Republic was also a victim of capitalism and ended its life in the adversity it was reared in. Yet many of the pieces that made up the Republic Steel Corporation survive, proving that the company's masters knew how to find the good raw materials needed for success.

. . .

The photographs in the Republic Steel section come from several sources. Photos from the 1930s and early 1940s are from an early collection of Republic photographs that the Western Reserve Historical Society acquired after the Republic-LTV merger, along with the previously noted Corrigan, McKinney construction photographs. This collection, containing several dozen Republic images, is a beautiful example of 1930s industrial photography at its peak. In most cases the photographers are unknown, though several photographs are credited to Charles Ackerman of Cleveland, a local photographer who did some corporate photography. However, only a handful of these photographs are credited to any photographer. Unfortunately the photographs had been stored a long time in a basement at Republic Steel, and upon unwrapping the boxes containing the photos, we discovered that the cellophane tape holding a caption sheet on the back of each photograph had slightly bleached out the following image in the stack. Many excellent photographs were damaged in this manner. However, quite a few wonderful photos were undamaged and selected for this section.

The remaining photographs, including the color photos, are from the Republic Steel negative files, taken in as part of the LTV Steel Collection. These images were used for a variety of publications, including the monthly *Republic Reports* magazine that went to employees and sales clients, advertising materials, company publications, corporate presentations, and internal newsletters. The images in this chapter cover the late 1940s to the 1960s, though the negative files go all the way to Republic's final days before the merger with LTV. For these photographs, the negative number is noted when possible. The color images are also from the same negative file, though they were organized differently, and all marked with a C prefix. Several good images came out of a drawer marked "Cleveland Engineering Department," and those images are noted with a negative number as well, if one appeared on the negative envelope. In nearly every case the photographer is not credited. However, Fred Perdue is credited as the Republic Steel photographer in decades' worth of *Republic Reports,* so it is plausible that he took nearly all of the photos from the negative files.

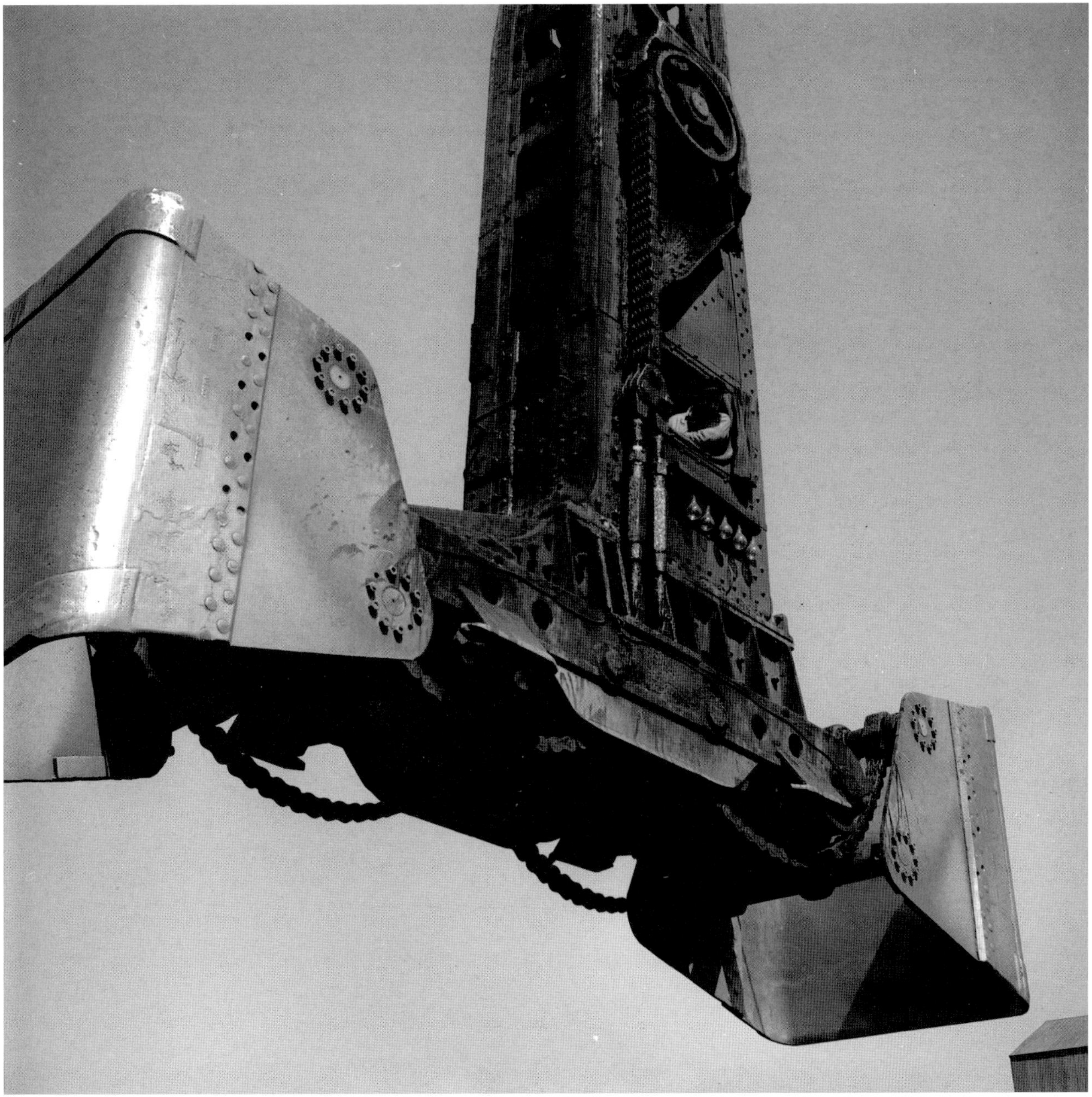

The giant Hulett ore unloader removes iron ore from the hold of the steamer *J. E. Upson* in this 1960 photo and places it into waiting rail cars. Iron ore for Republic's mills came via bulk carriers across the Great Lakes from Minnesota or Upper Michigan. (photo 4230C)

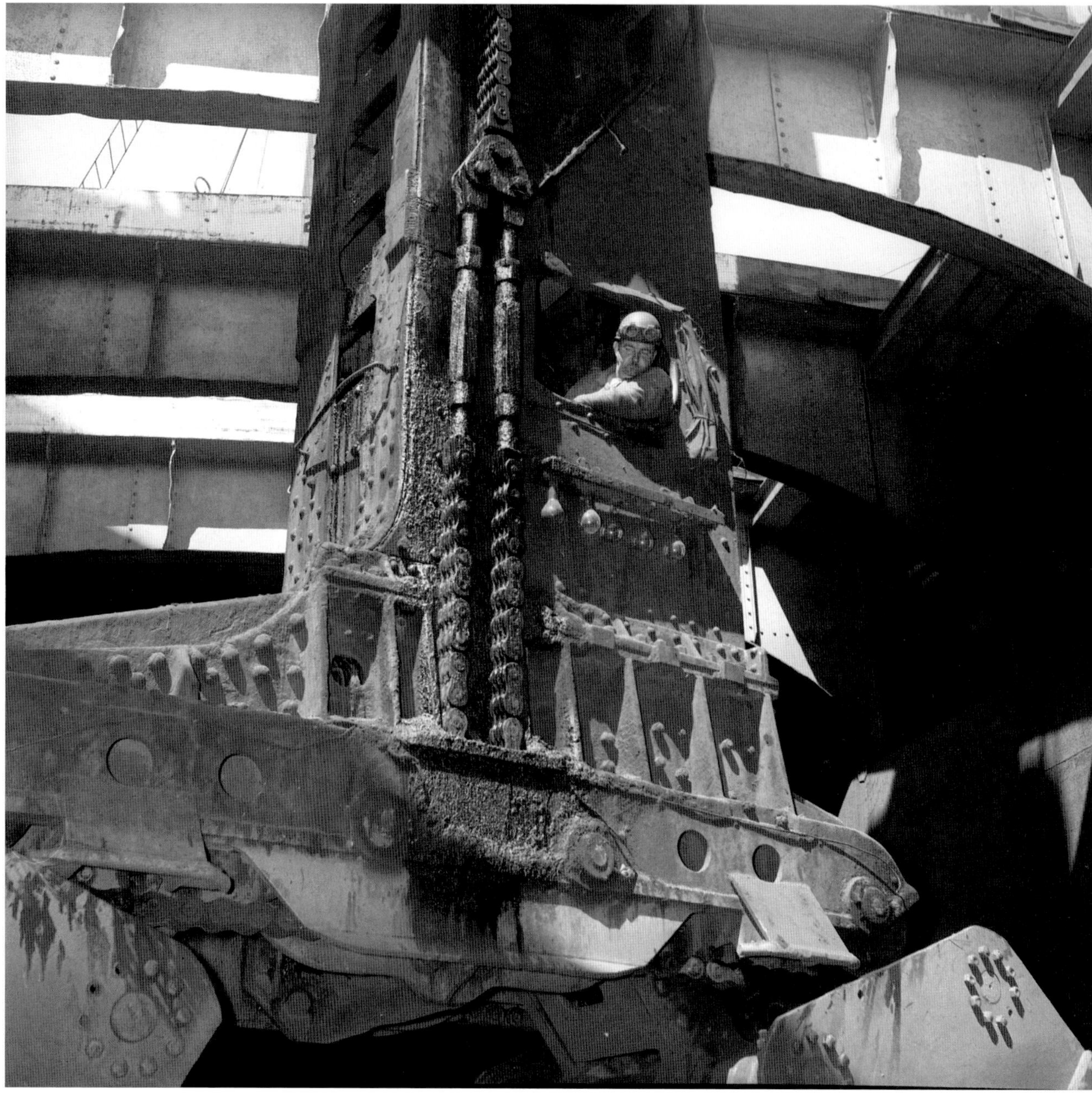

Unlike other types of bulk-moving devices, the Hulett operator rides in the leg, which was lowered into the ship's hold. This 1960 photo shows how the operator could navigate the unloader into the hold's narrow hatches. (photo 4231G)

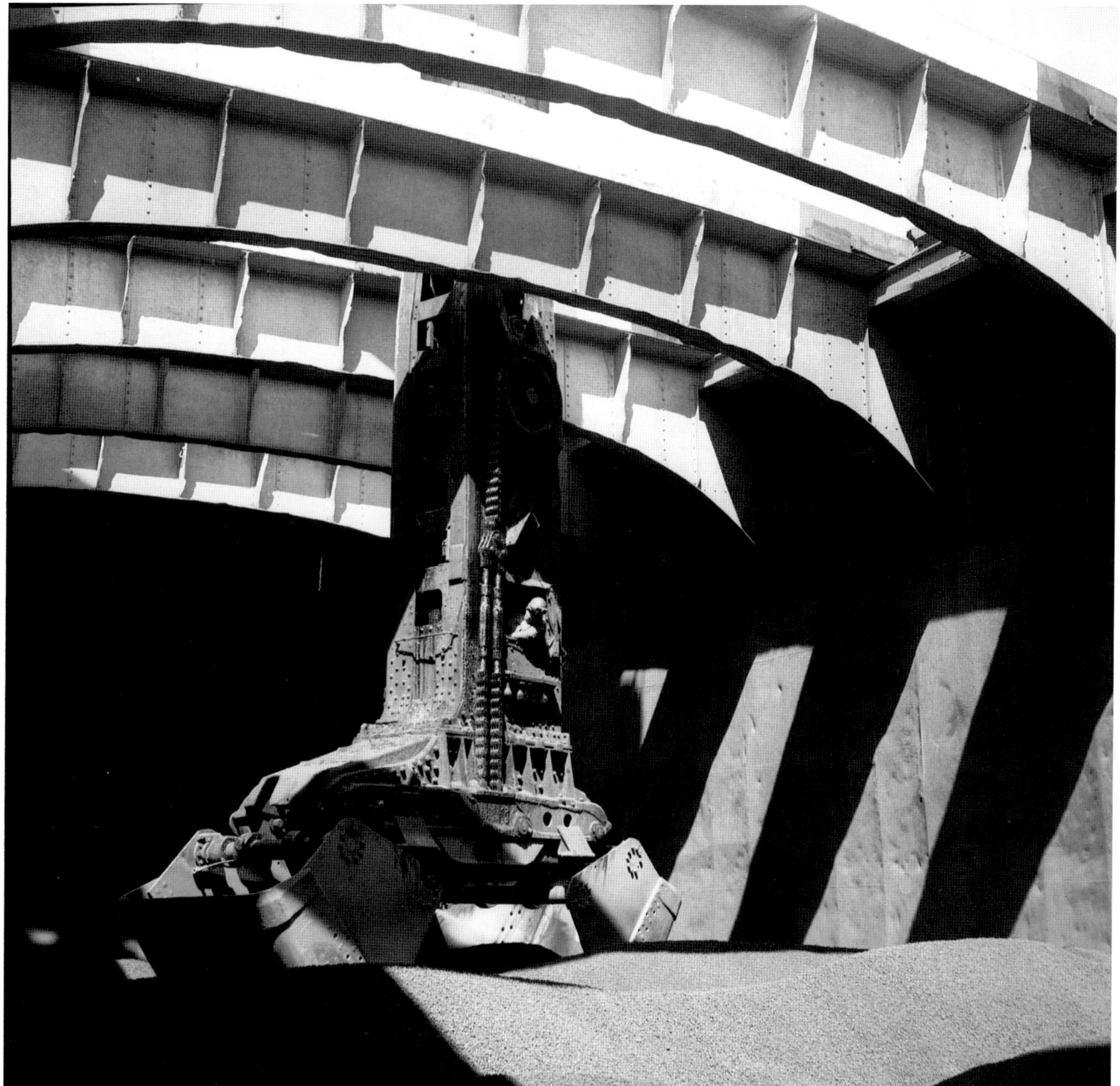

Once the Hulett unloader's leg landed on the iron ore, the giant buckets closed, pulling in as much as 15 tons in a single bite, as in this 1960 photo. The Hulett unloader leg then rose out of the hold to deposit the ore into waiting rail cars. (photo 4231H)

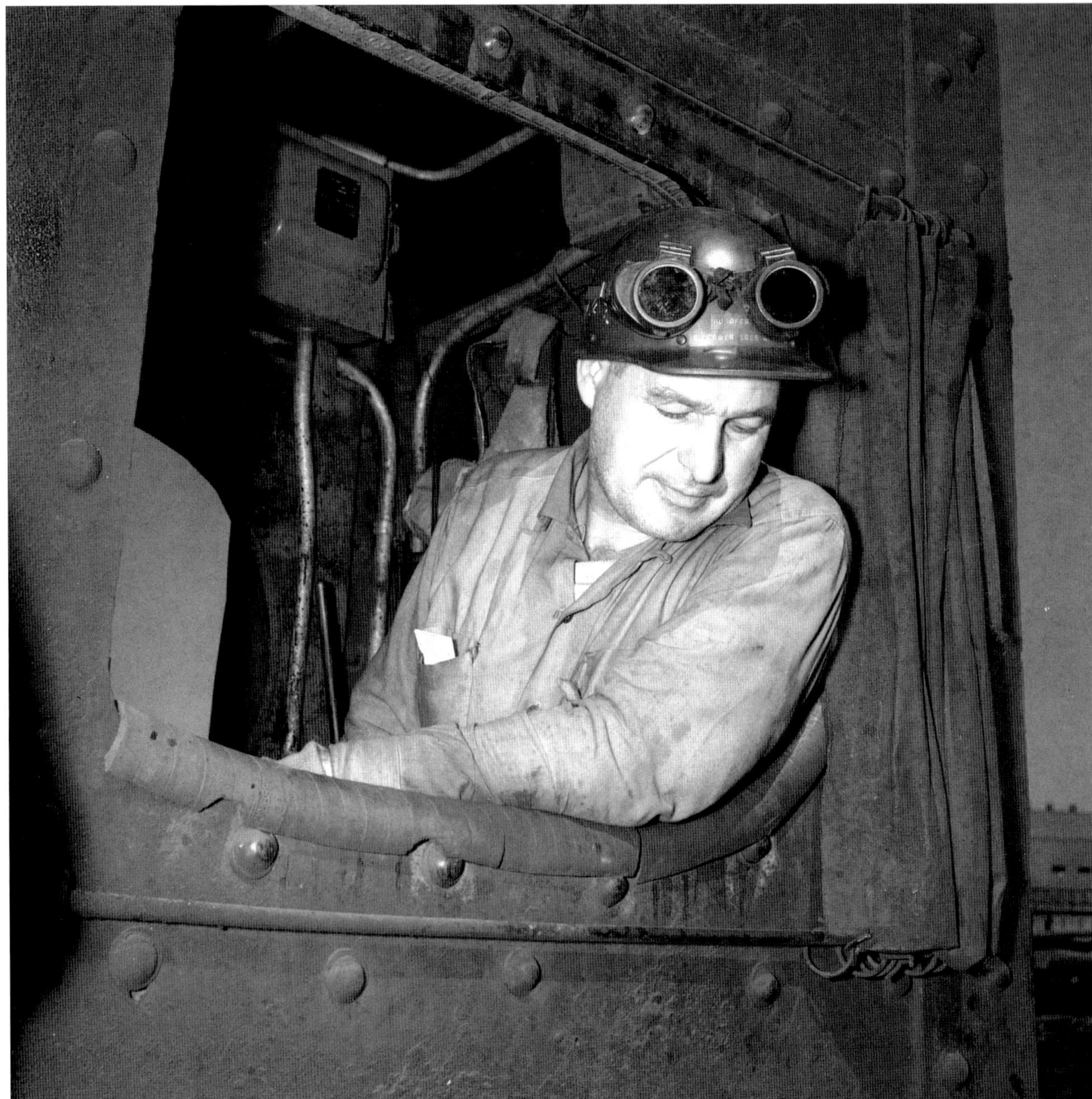

In this 1960 photo Hulett operator T. Durbin leans out of his cab in the leg of an unloader to watch where he's going. Operators often had to stand in their cabs for hours while they unloaded a ship's hold. (photo 4230A)

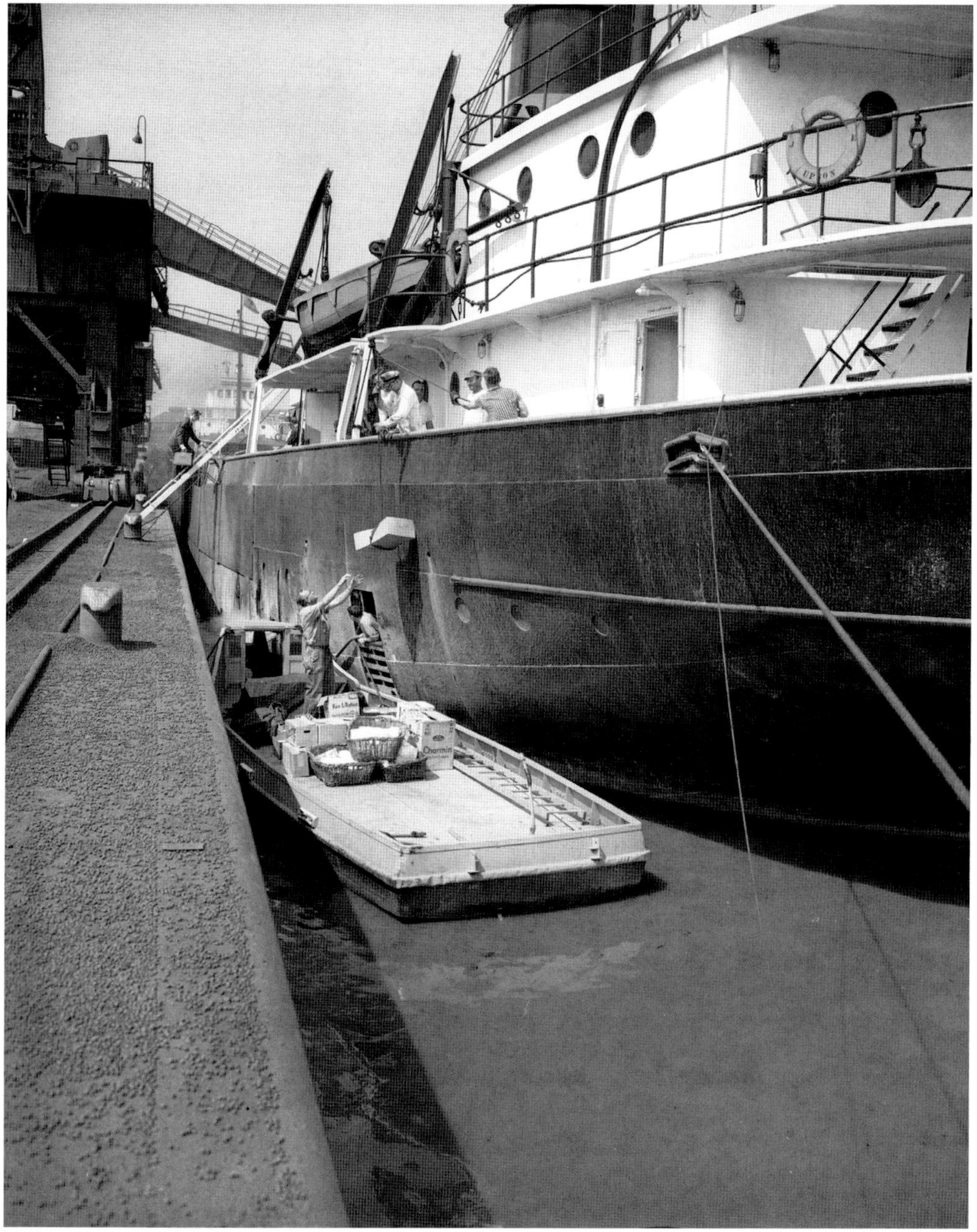

While the vessel is tied up at the ore dock for Blast Furnaces 5 and 6 in Cleveland, the Hulett ore unloaders attend to their task in the background, and the crew of the steamer *J. E. Upson* takes on supplies. (photo 4263)

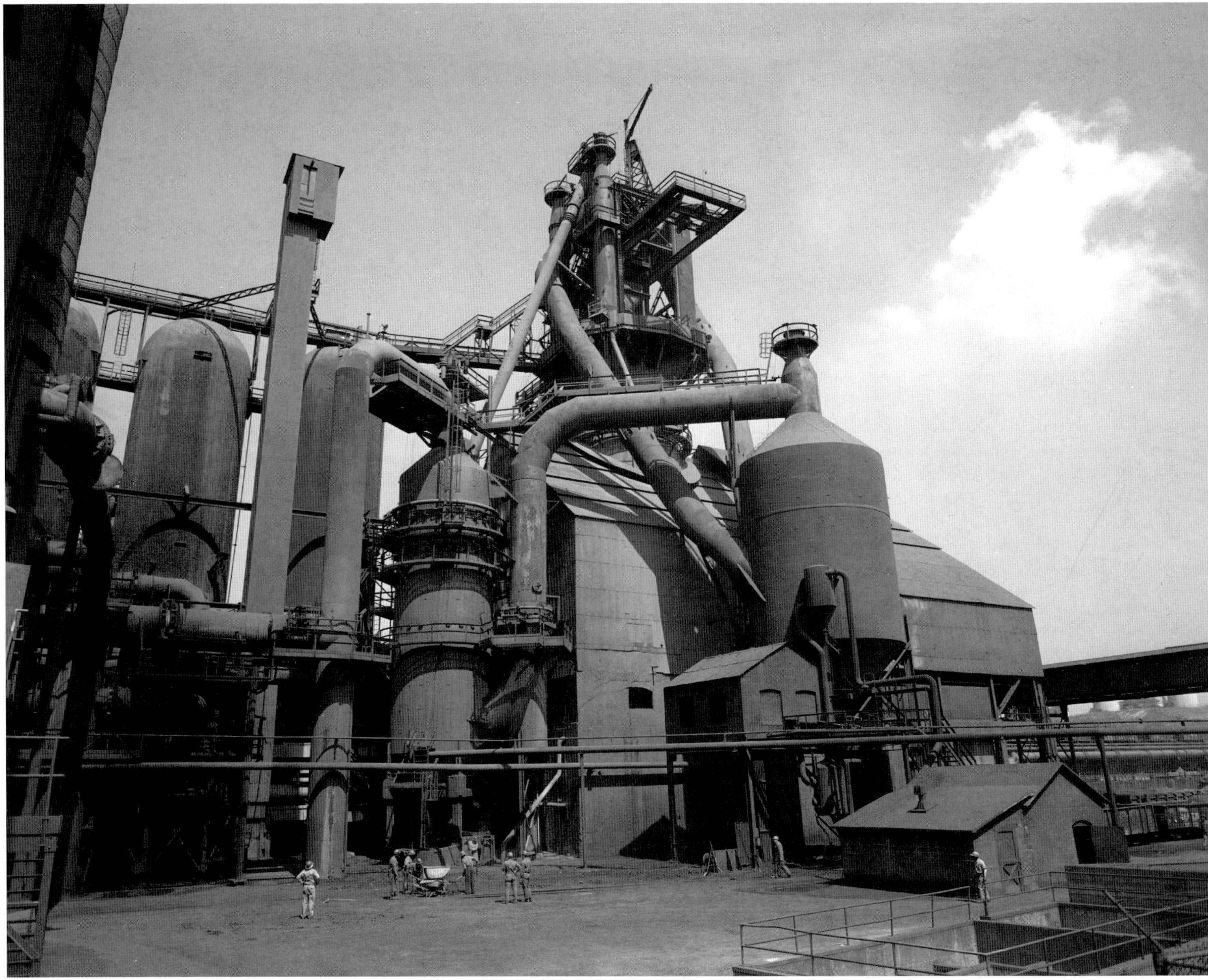

Once the ore was unloaded from the ship, it went into the blast furnace, where it was refined into pig iron. The scale of the furnaces can be seen in this 1962 photo, where Cleveland Blast Furnace 5 dwarfs the work crew at its base. (photo 5004C)

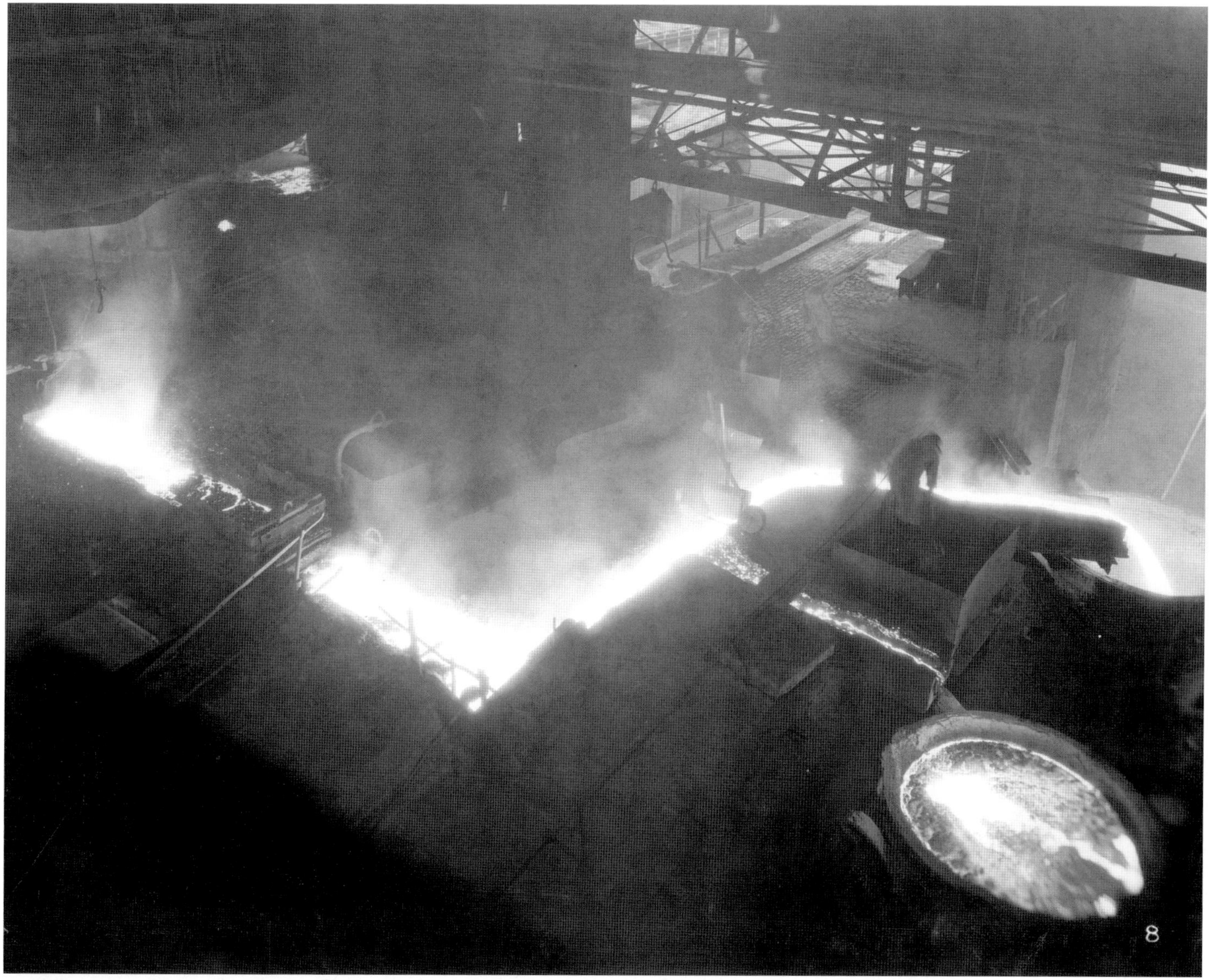

Casting a heat of fiery metal from the huge blast furnace. About four or five casts, each averaging 150 tons of pig iron, would be made in one day in the mid-1930s, when this photo was taken at Republic's Canton, Ohio, mill. Note the precarious position of the worker as a hot stream of molten pig iron pours into a ladle to his right and a stream of molten slag pours into a slag thimble to his left.

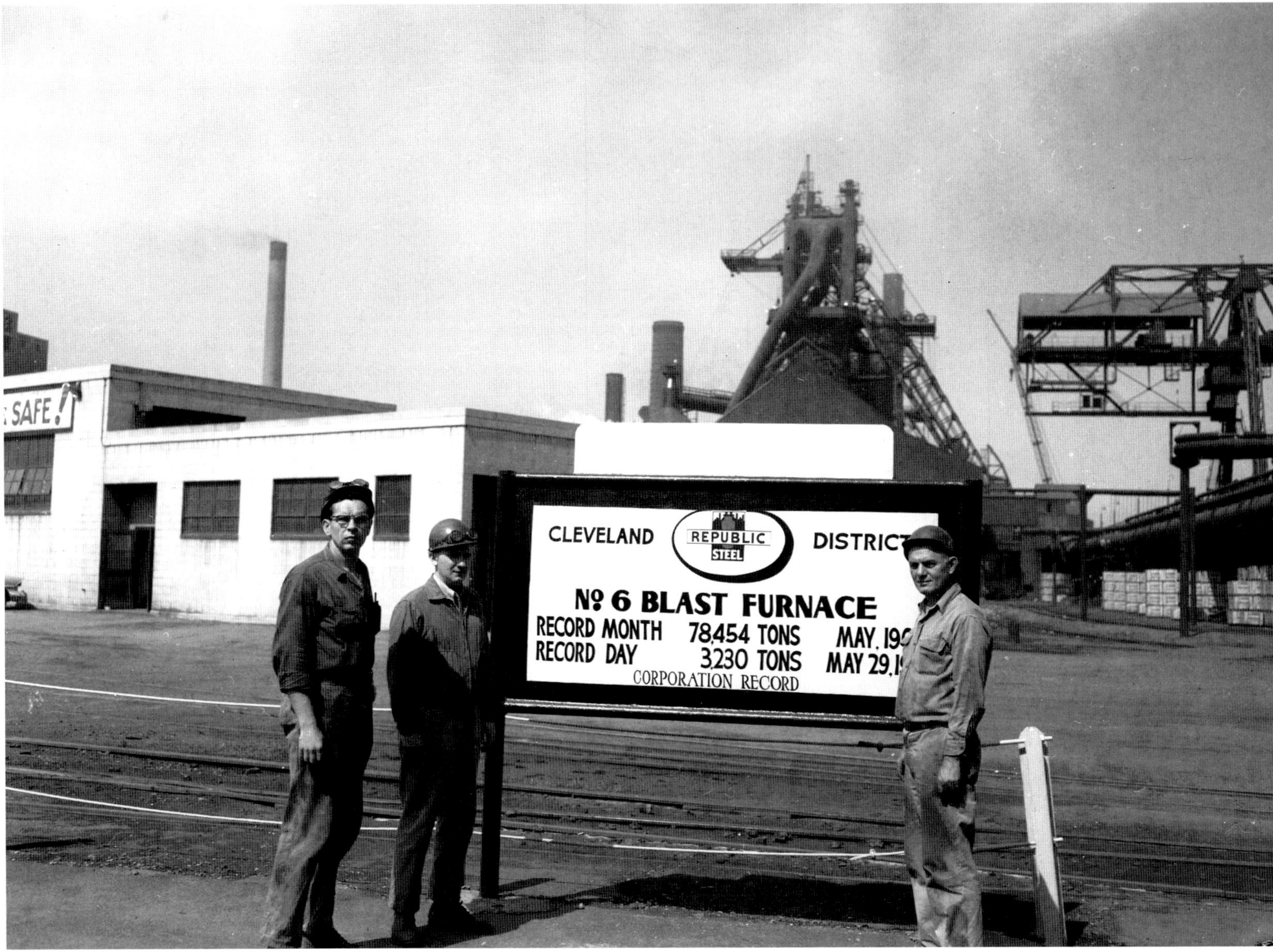

The capacity of Republic's blast furnaces increased over the years, as shown in this sign noting the record for Cleveland's Blast Furnace 6 was 3,320 tons on May 29, 1961. This capacity was much greater than the furnaces of the 1930s, as in the previous photo. (Cleveland Engineering Department Negative 9)

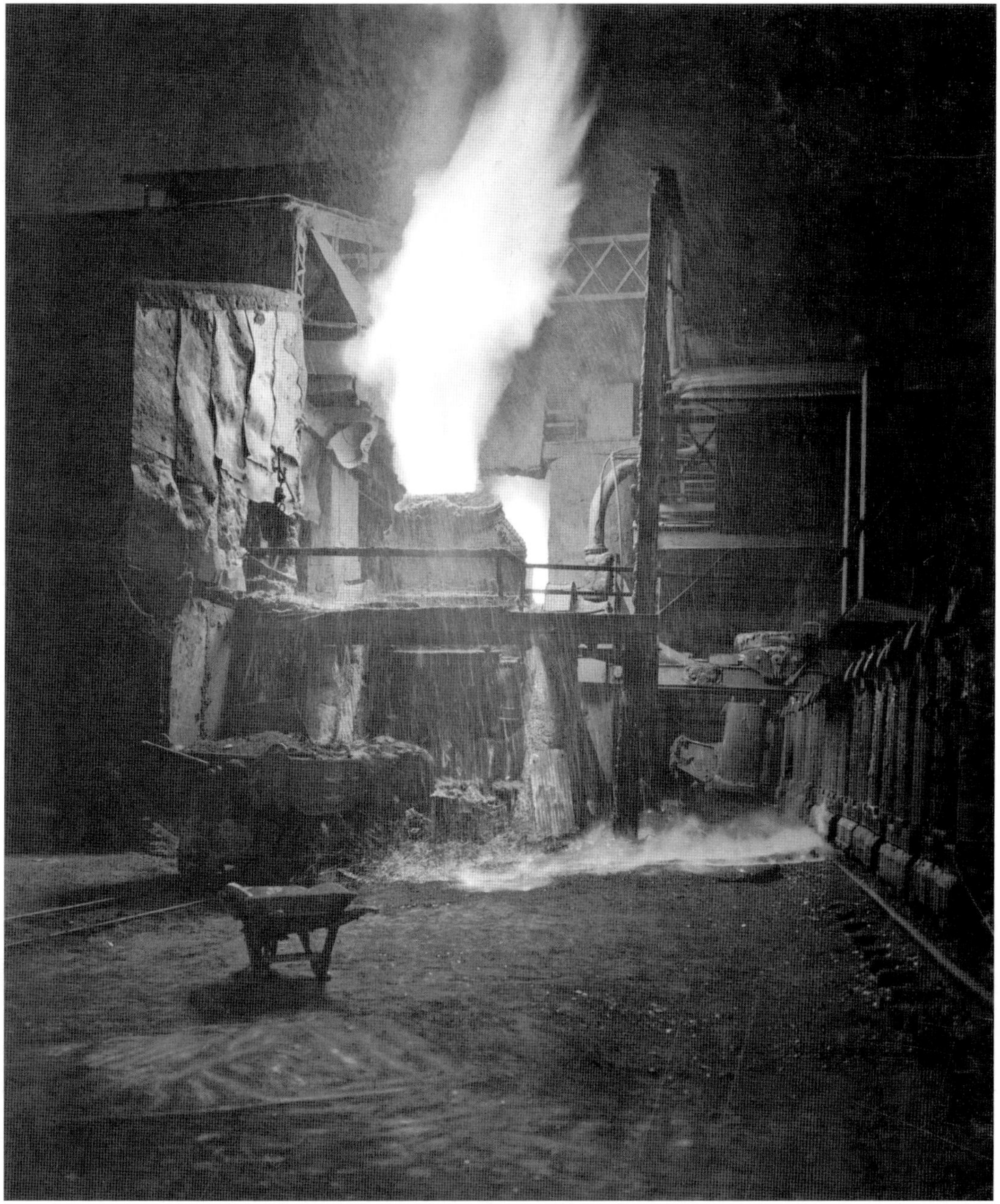

This photo from the late 1930s shows the Bessemer plant in operation at night at Republic's Youngstown mill. The converter tilts upward as a blast of air is blown through the hot metal.

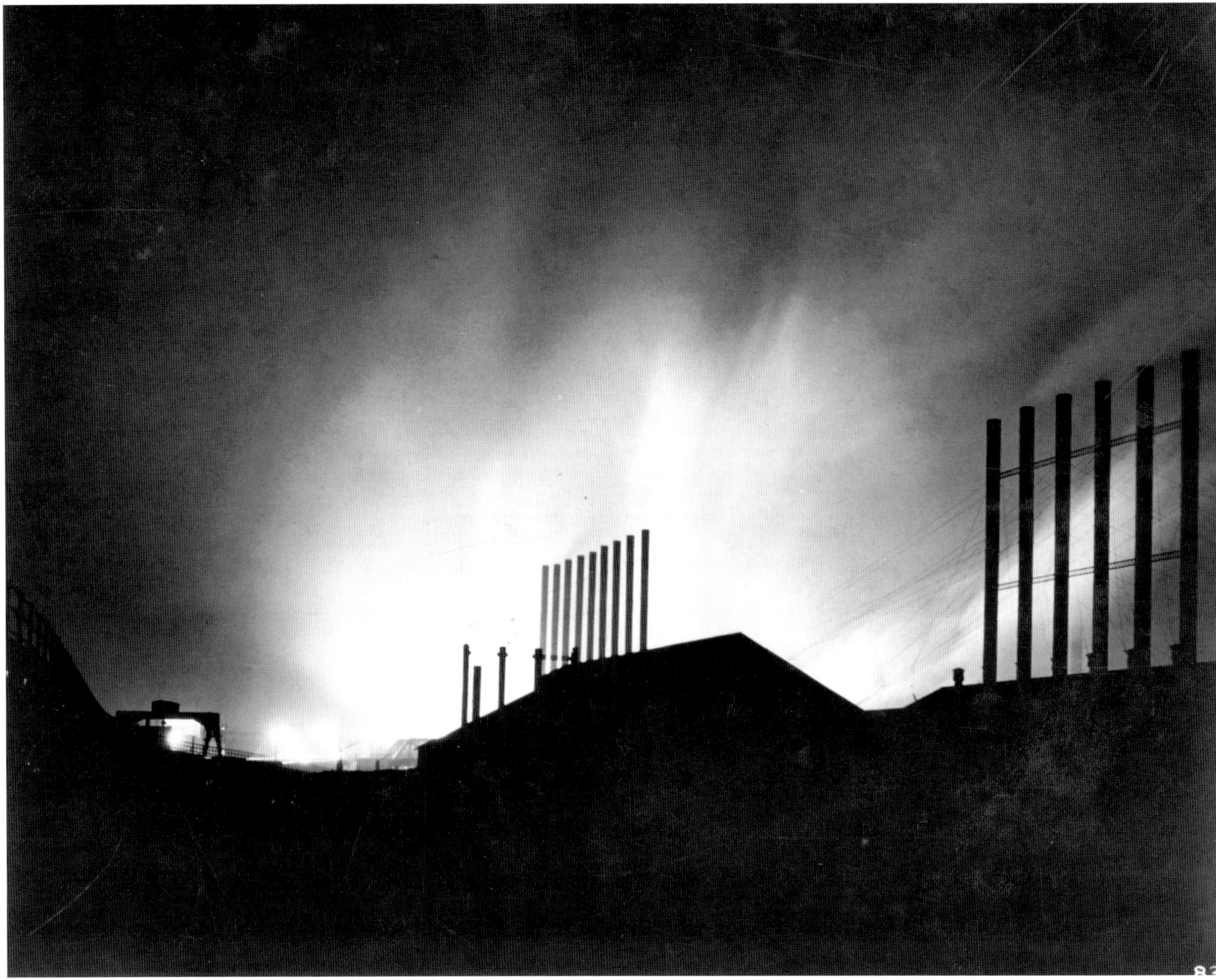

Bessemers in full blow were an impressive sight. In this nighttime shot the Bessemers cast a glow over the Youngstown mill.

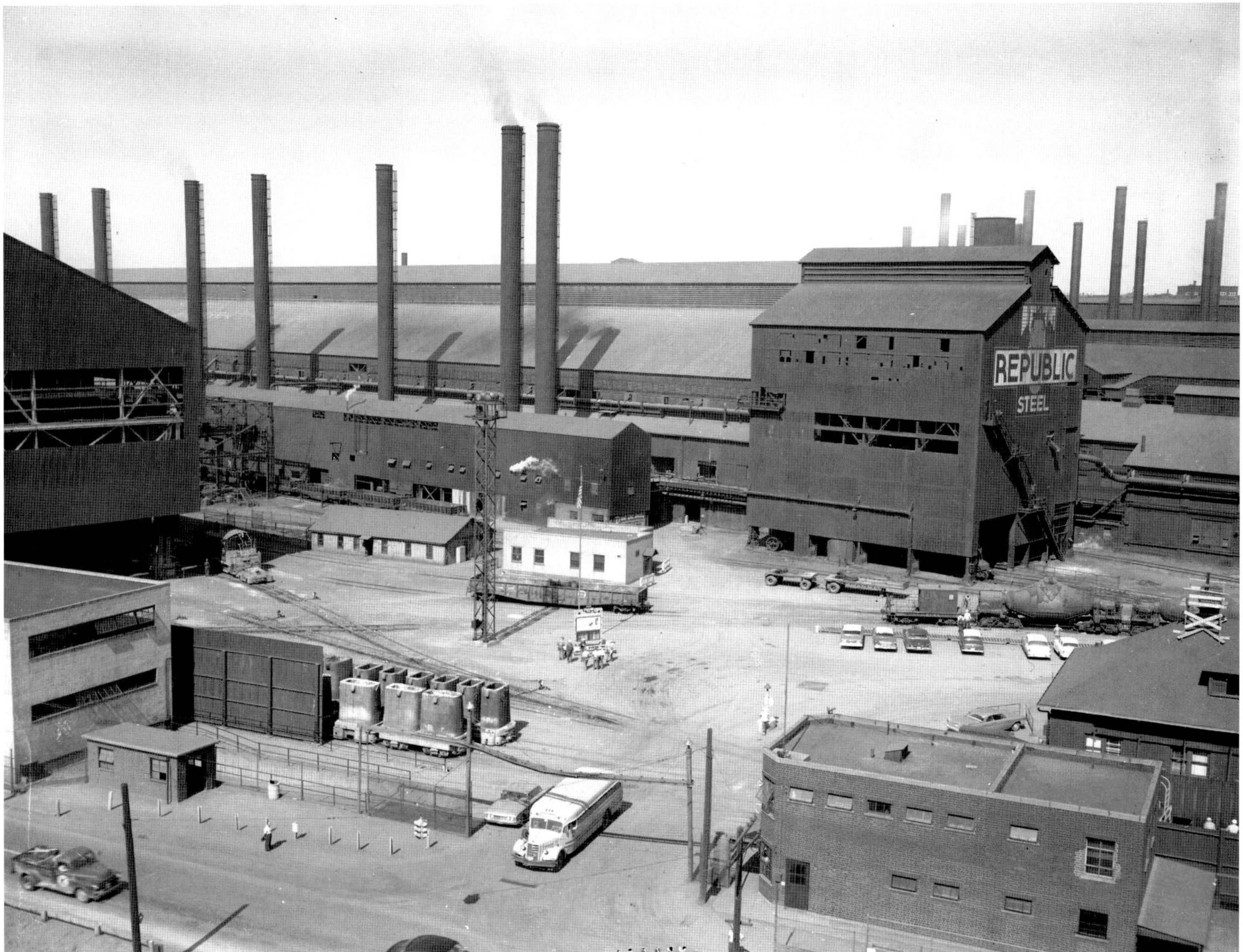

The Bessemers, while fast, were often not precise enough for all the varied grades of steel required by customers, so they were soon surpassed by open hearth furnaces, which were slower, but could be controlled more precisely. Open hearths dominated steelmaking into the late 1960s until basic oxygen furnaces were introduced. This photo shows the gate at the open hearth department of the Cleveland mill in the early 1960s.

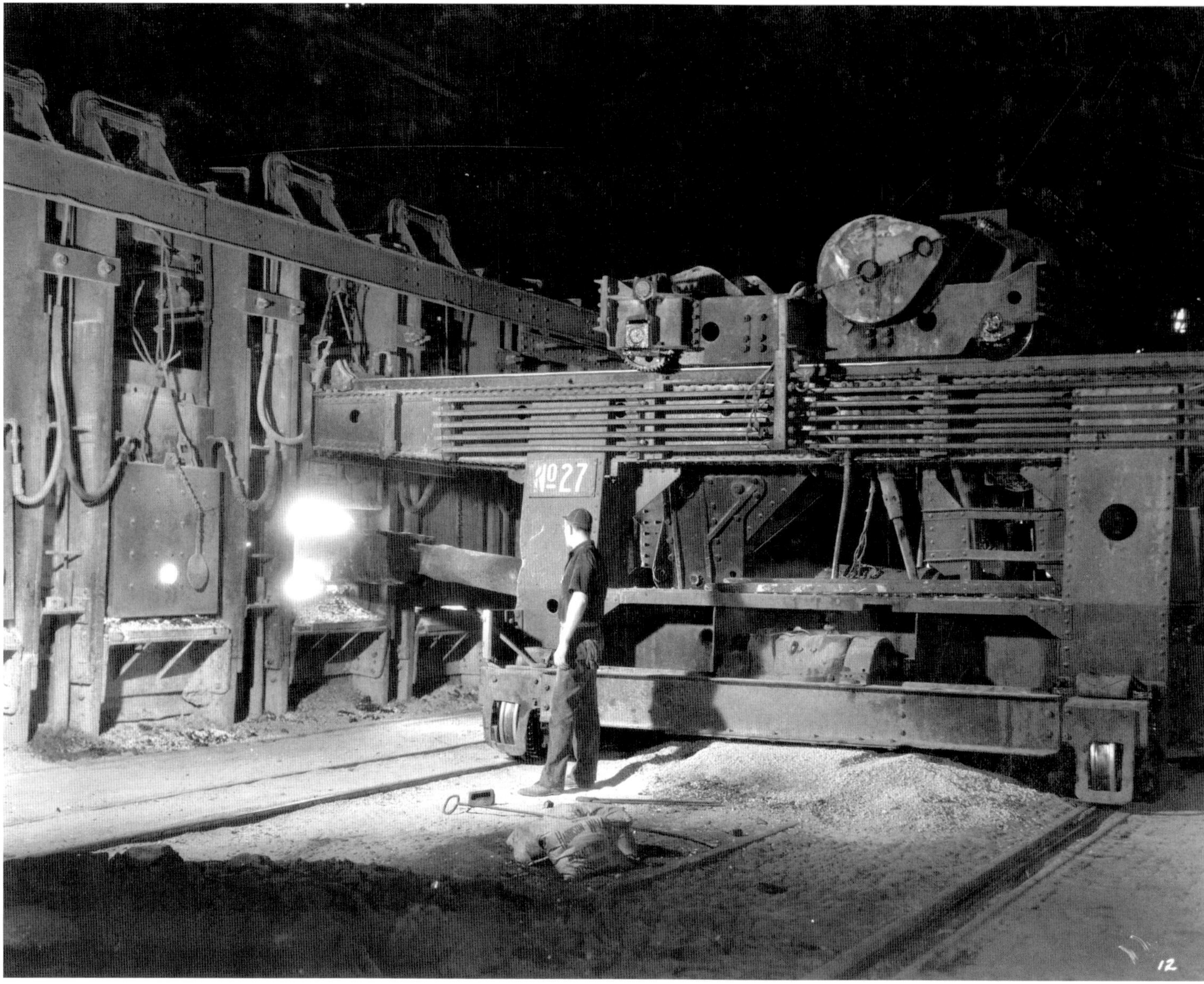

Workers had to charge open hearth furnaces with scrap metal. They usually placed the scrap in a large metal container, and a machine, pictured at the right, would insert the box into the hot furnace and turn it upside down to empty the scrap. This photo from the late 1930s was taken in Republic's Massillon, Ohio, plant.

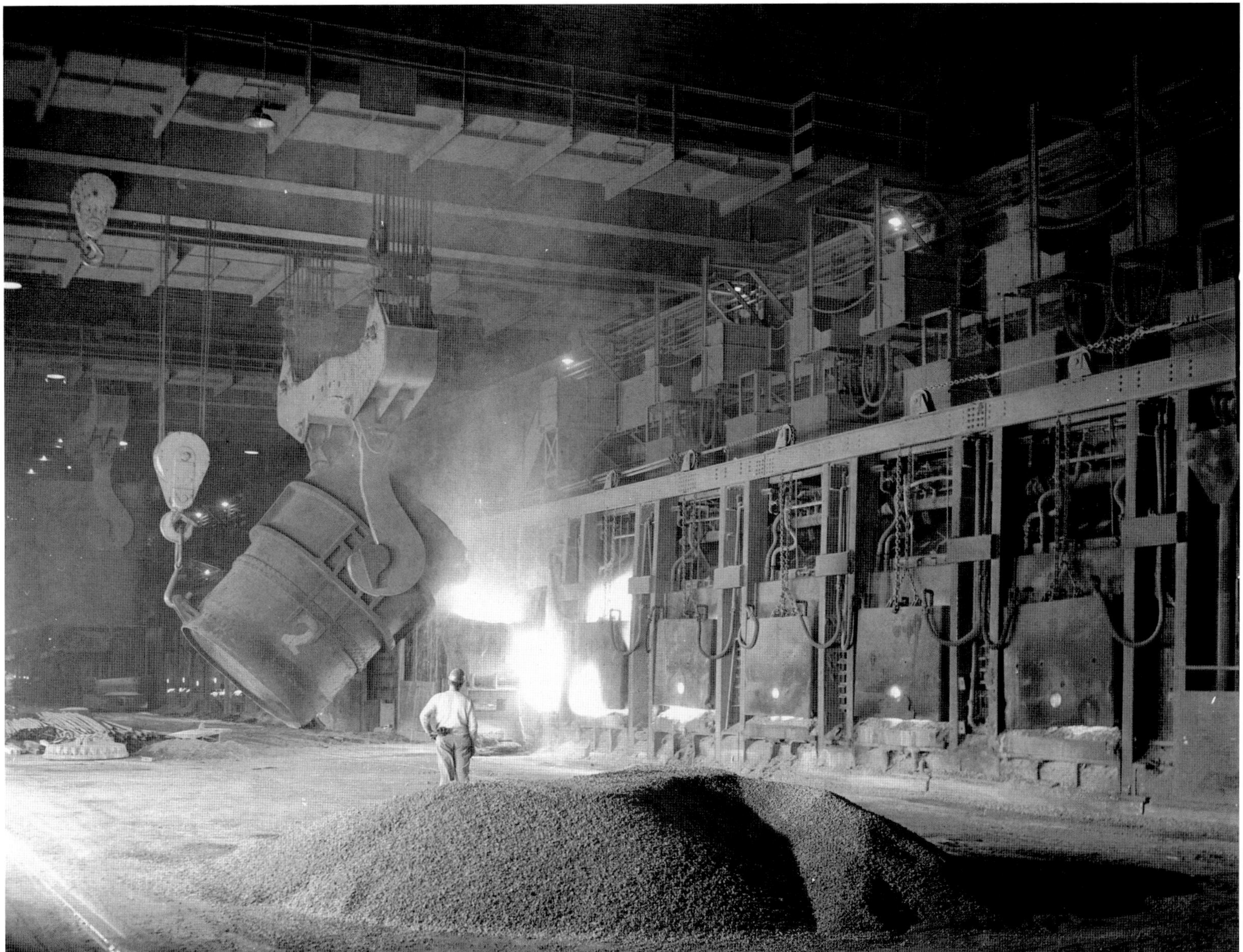

After workers charged the open hearth with scrap metal, they added hot pig iron from the blast furnace, as seen in this photo from Cleveland in the early 1960s. (photo C220)

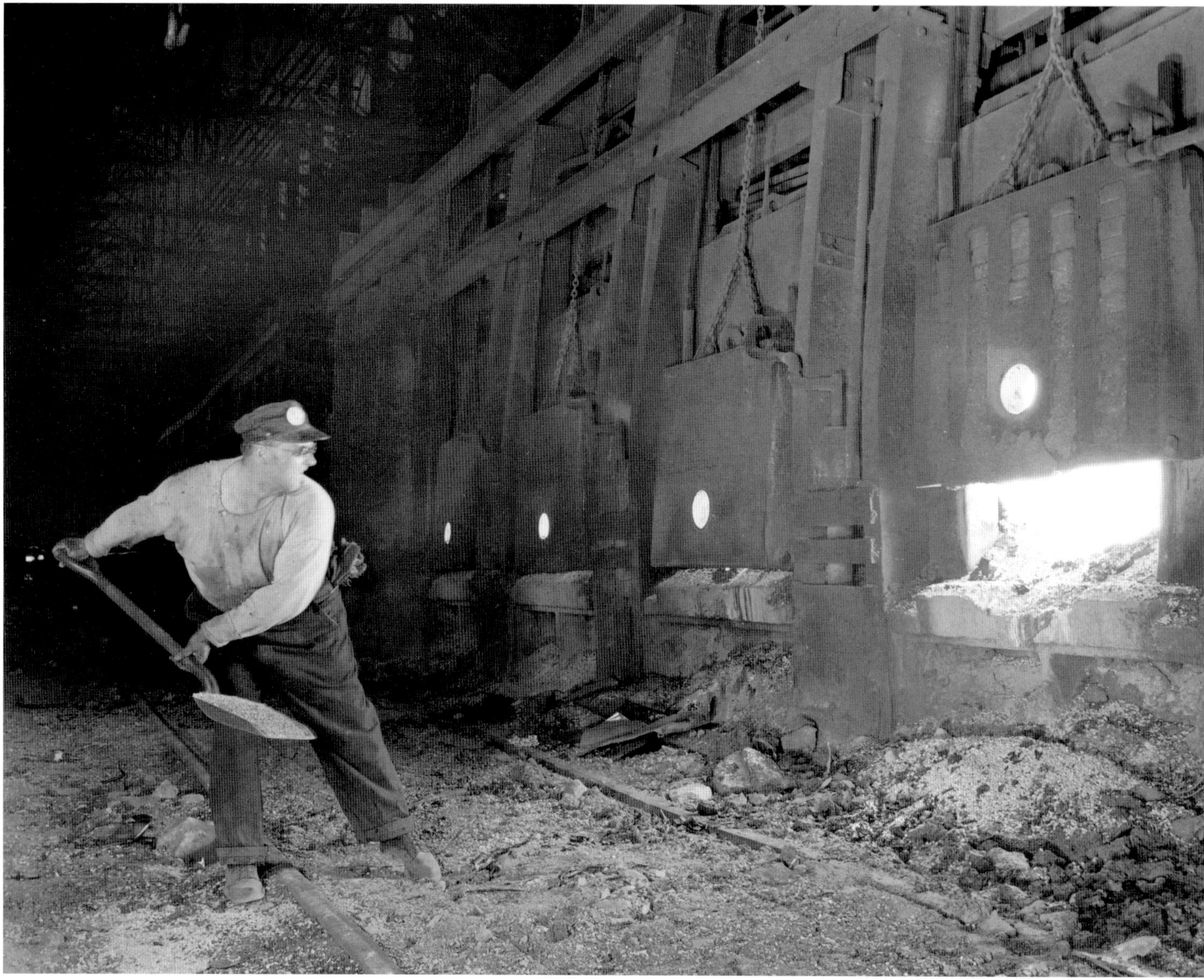

While the scrap and molten pig iron cooked in the open hearth furnace, workers shoveled dolomite into the furnace to protect the refractory material lining the walls. This photo is from the 1930s, when Cleveland's Republic plant was still called the Corrigan McKinney Works. Photo by C. W. Ackerman Studio, Cleveland.

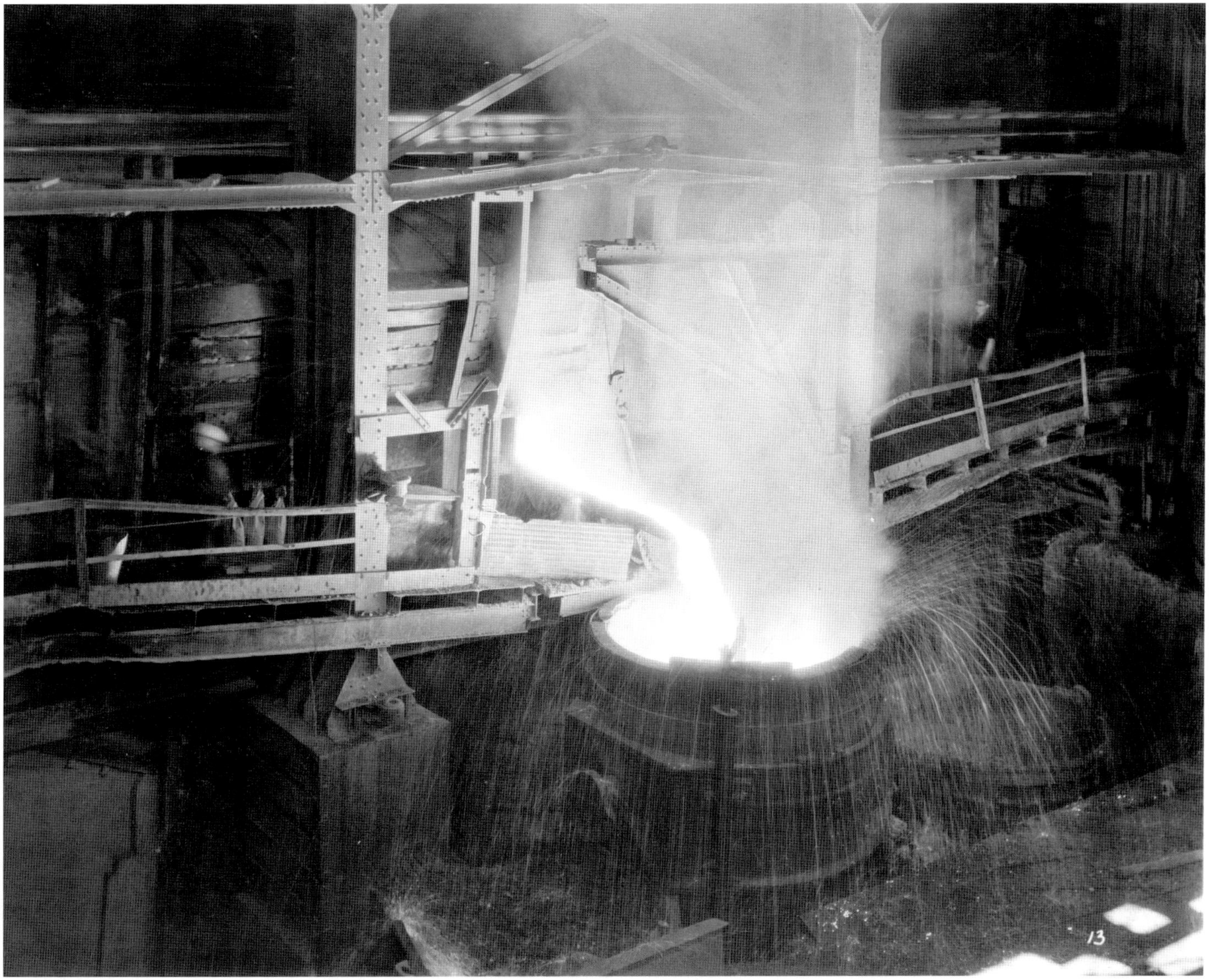

When the steel in the open hearth was judged to be ready, the rear of the furnace was tapped to allow the molten steel to pour out. During the late 1930s, when this picture was taken, an average heat was about 90 tons and required 10 to 14 hours to make, depending on the type of steel. Workers standing on the balcony behind the furnace could throw in alloys during the tapping. Note the blurry figure of the worker on the catwalk to the left. This photo was taken at the Massillon plant.

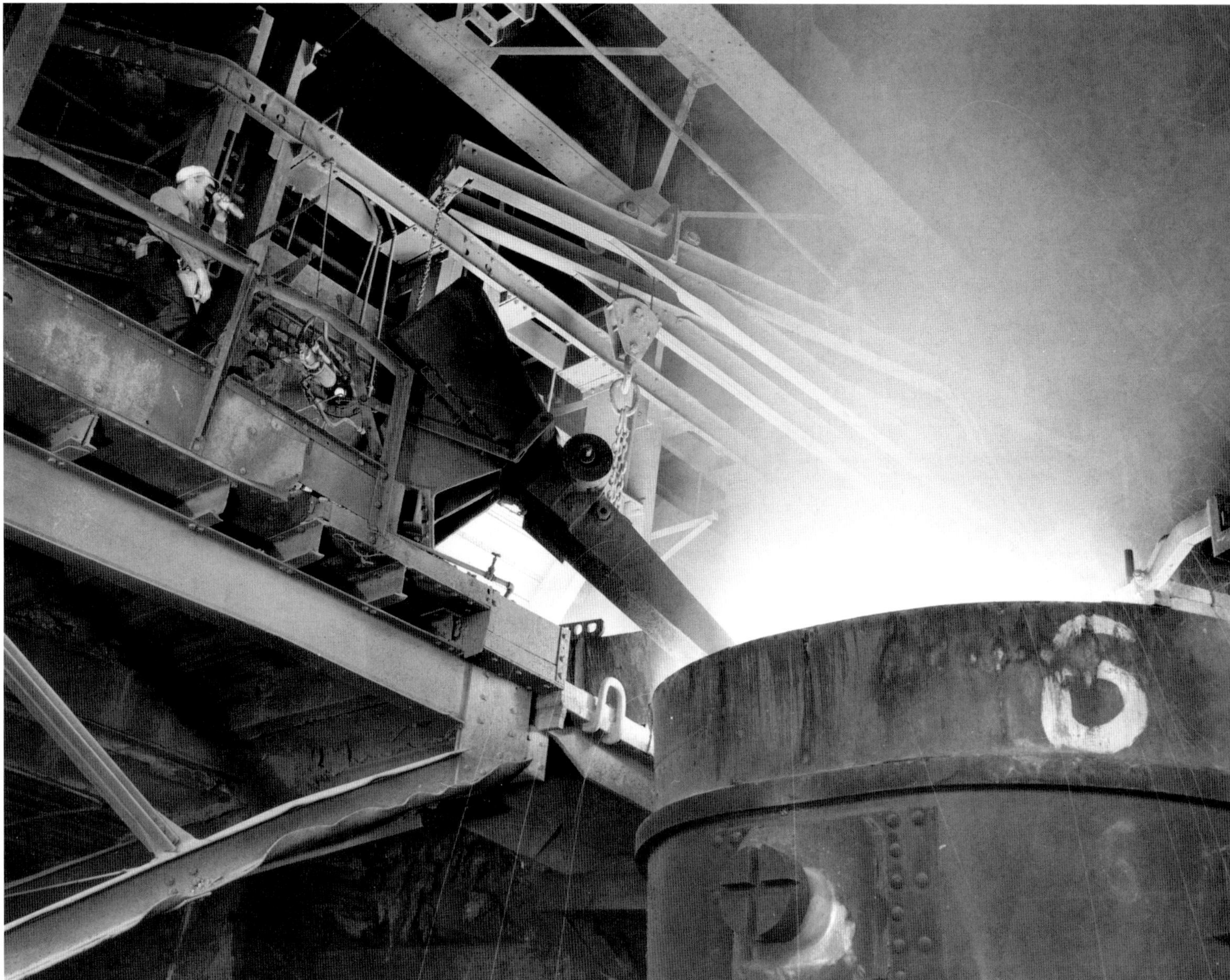

A worker observes an open hearth tapping at Cleveland's Corrigan McKinney Works in the late 1930s. Photo by C. W. Ackerman, Cleveland.

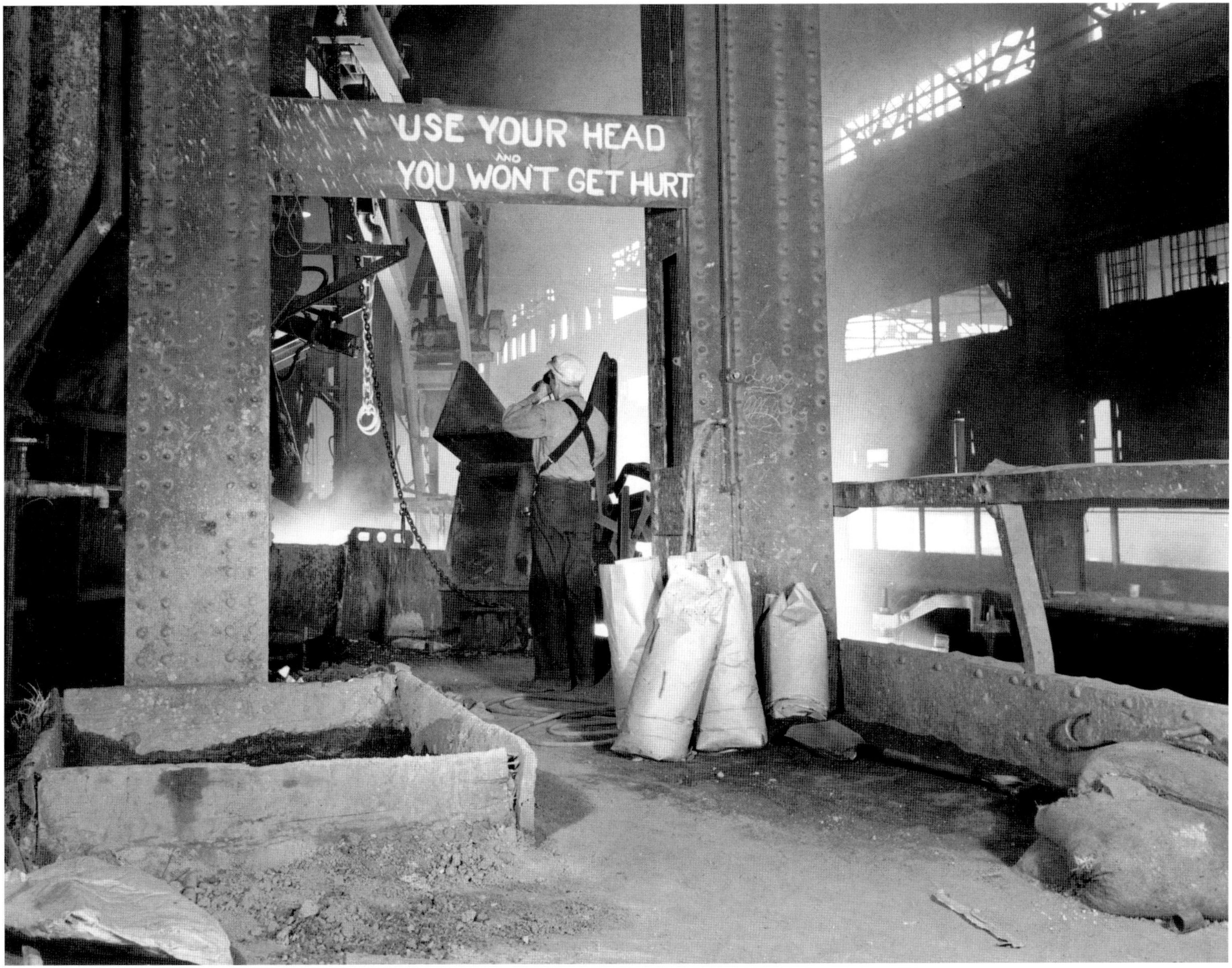

Definitely good advice at any mill. A different perspective of the worker in the previous photo, as he observes the ladle of molten steel. Note the bags, which probably contain alloy materials. Photo by C. W. Ackerman, Cleveland.

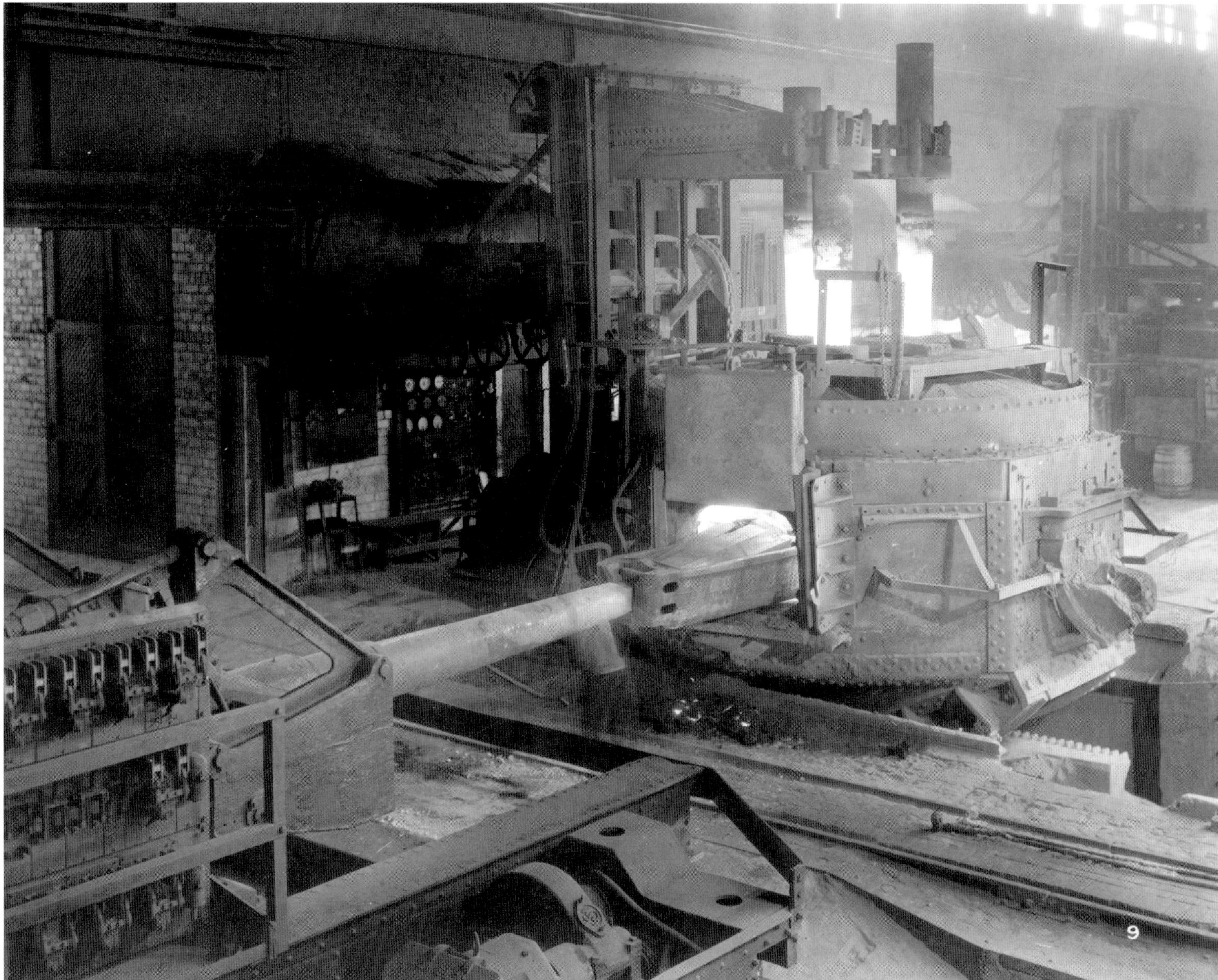

This electric furnace in the Canton plant is being charged with scrap metal, much like an open hearth furnace was charged.

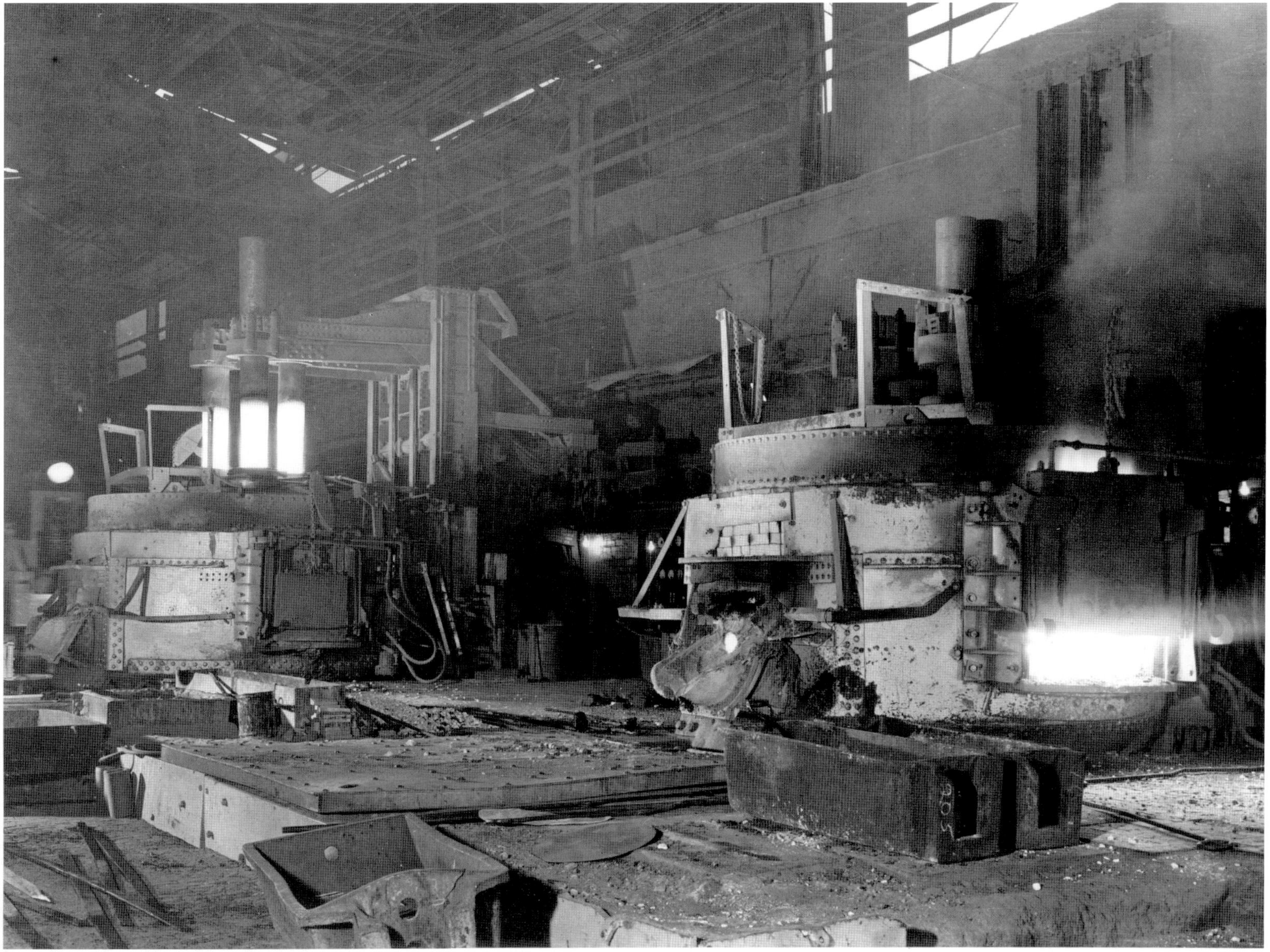

Two electric furnaces at work. The furnace on the right has its electrodes completely inserted, while the other is in the process of inserting them. The open door on the right furnace allows slag to be removed. Unknown mill, ca. 1940s. Photo by C. W. Ackerman, Cleveland.

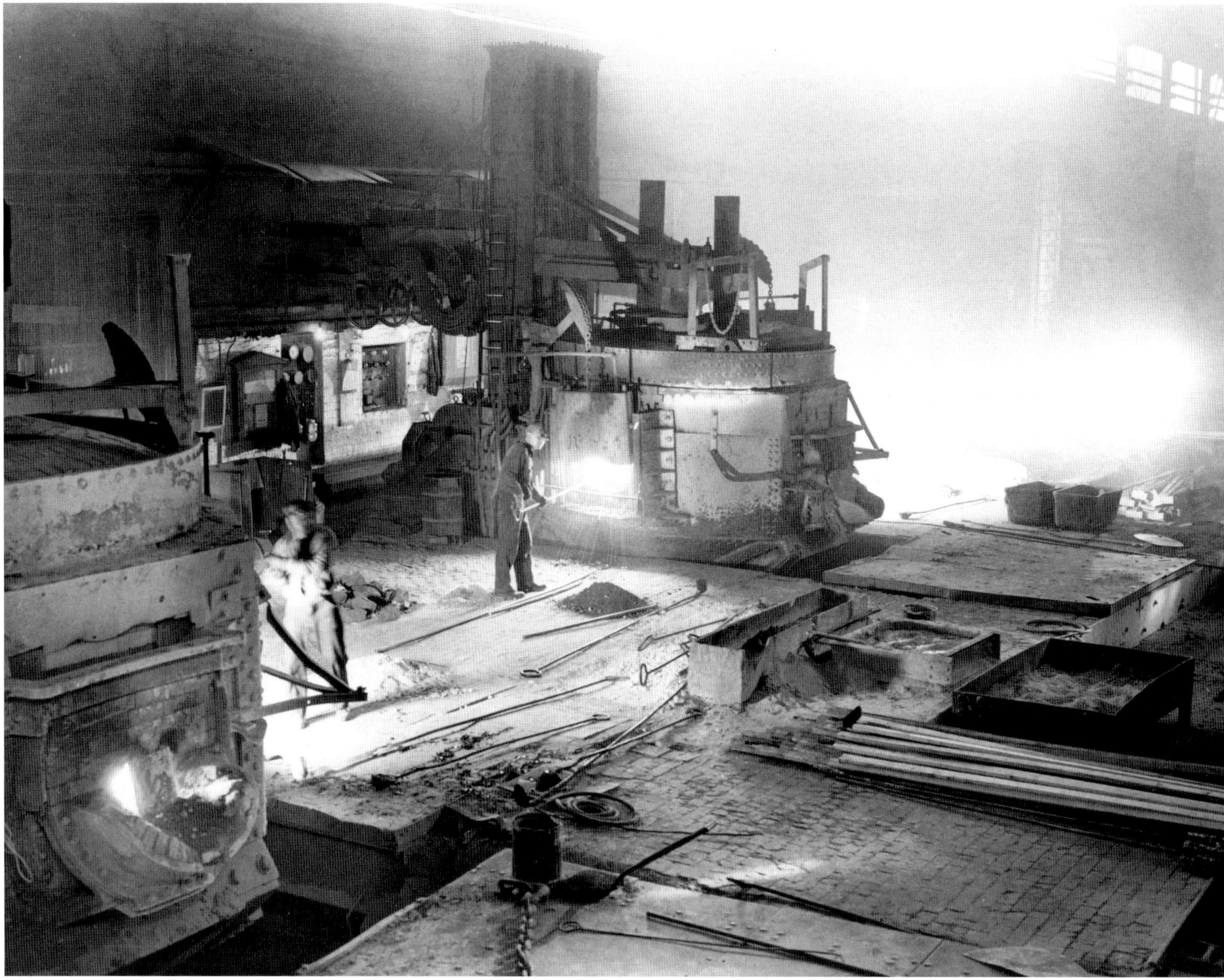

Workers probe the electric furnaces, possibly checking temperature or retrieving samples of molten metal for metallurgical tests. The workers, neither wearing safety gear, stand only a few feet from the open doors and the molten steel within. Unknown mill, ca. 1940s. Photo by C. W. Ackerman, Cleveland.

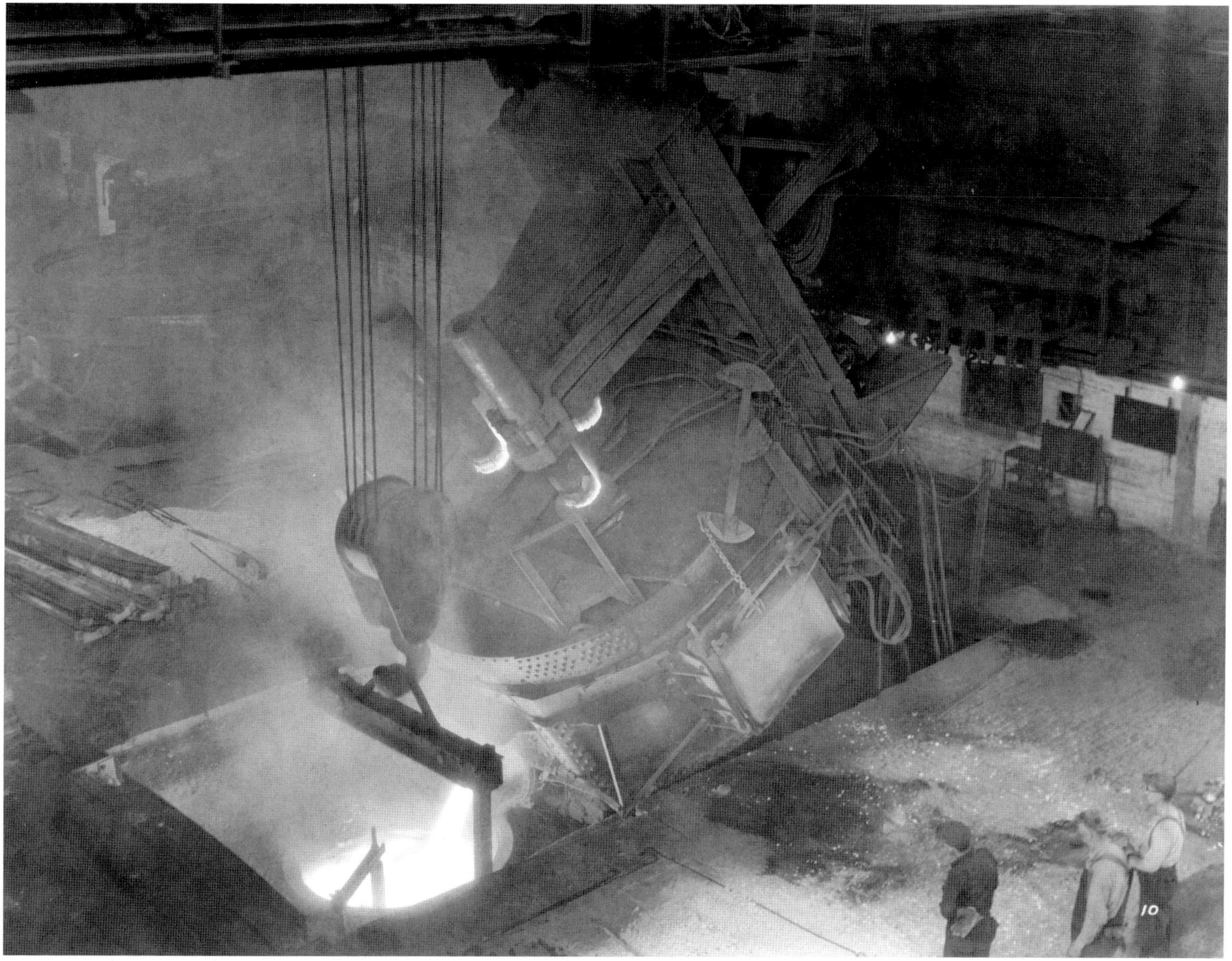

Tapping one of a battery of electric furnaces in the Canton plant of Republic's Central Alloy Division. In the Canton mill, furnaces like this one made Republic's fine alloy steels, such as Enduro stainless steel. An average heat is about fifteen tons. Photo ca. late 1930s.

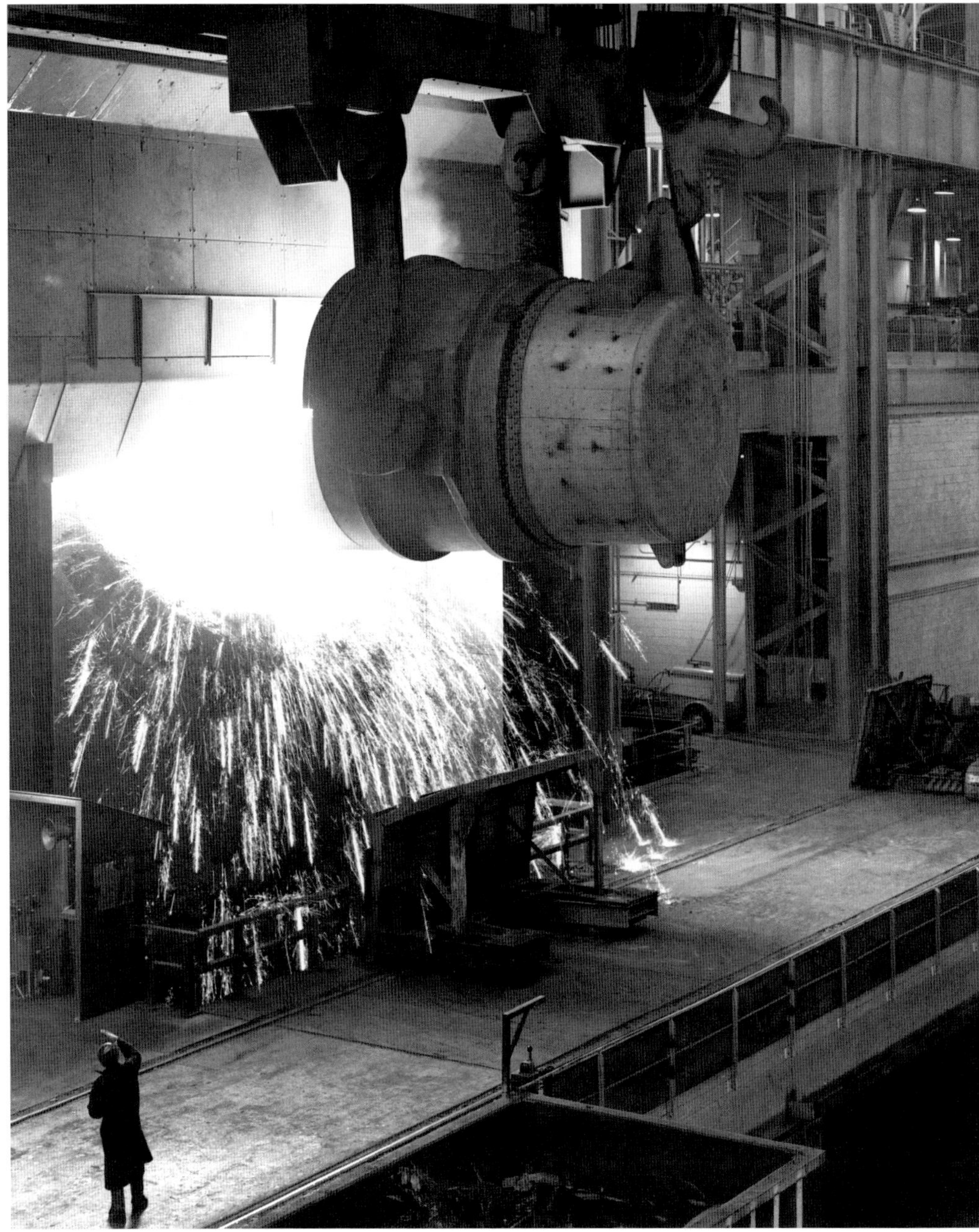

By the mid- to late 1960s, basic oxygen furnaces (BOF) were supplanting open hearth furnaces. Like a Bessemer, a BOF vessel tilted down, was charged with scrap and hot metal (as seen here in the Cleveland BOF in 1966), and then tilted back upward. An oxygen lance was lowered in and pure oxygen was blown in at supersonic speeds. This fueled the furnace and mixed the contents so that a complete batch of steel was made in thirty minutes. (photo 6226J)

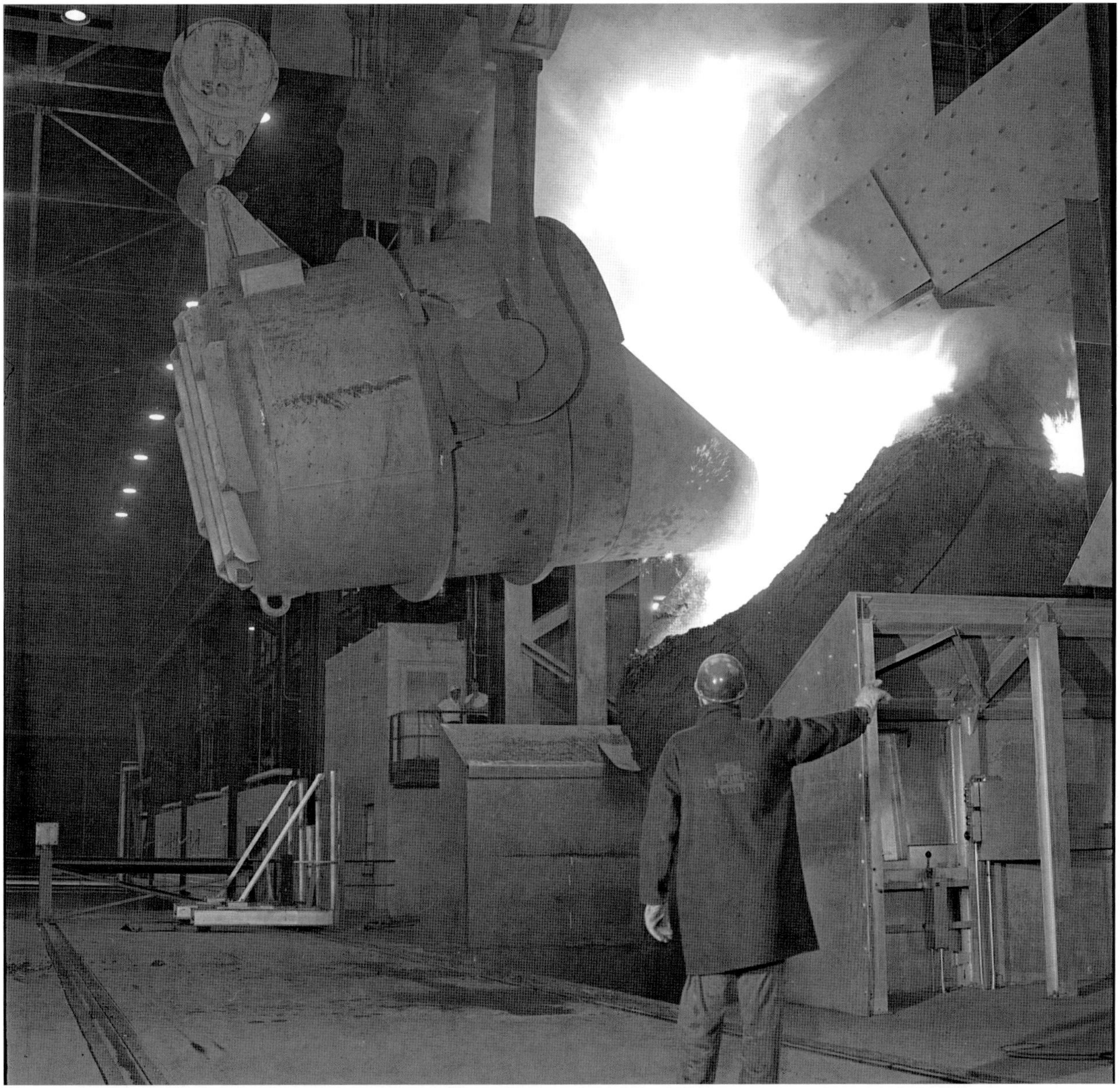

Charging the basic oxygen furnace, as seen in Republic's Warren mill in 1965, is the most visually impressive aspect of current steelmaking practices. (photo 5833K)

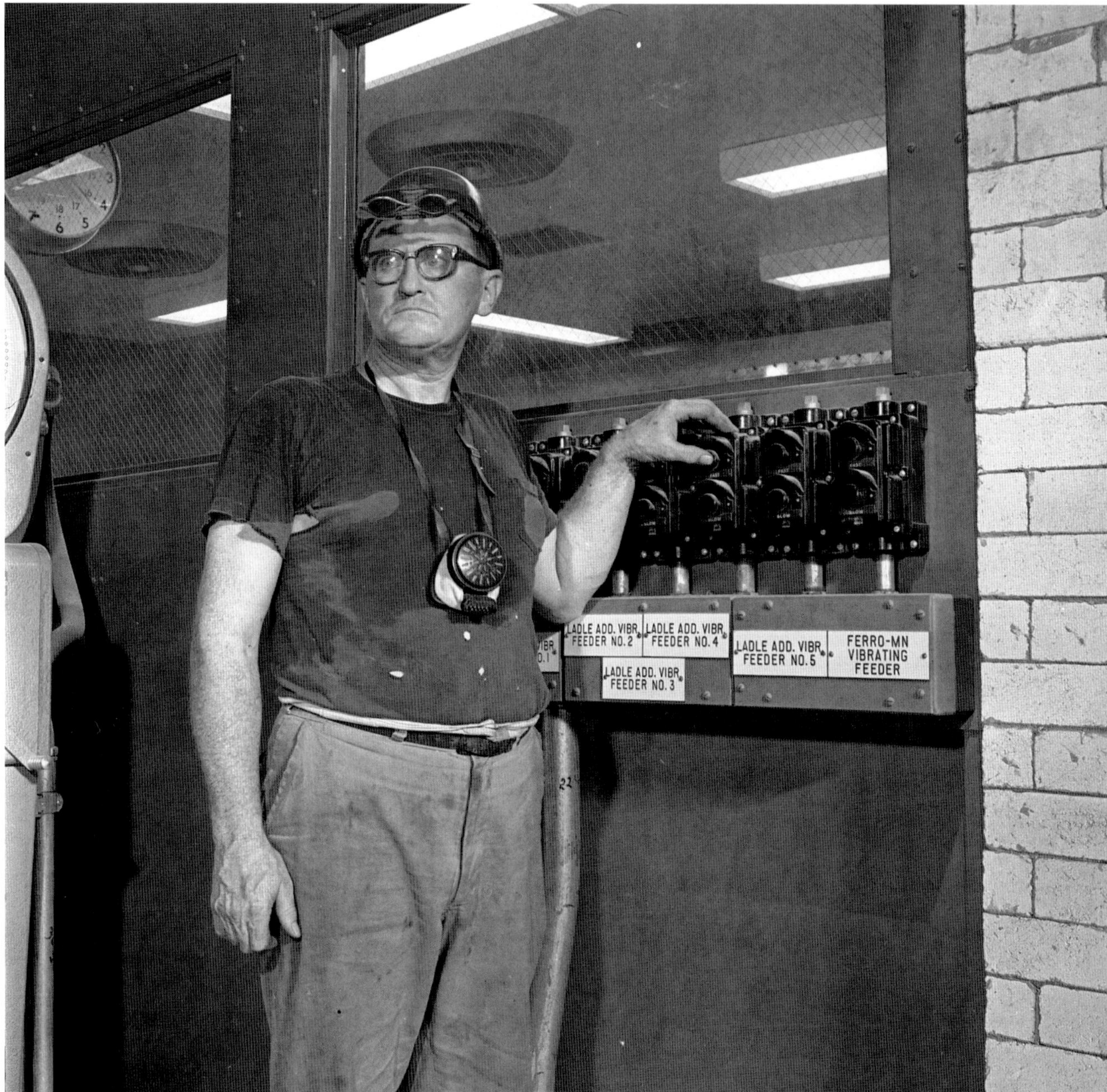

Earl King, second vesselman on the Warren crew, watches his basic oxygen furnace in 1965. (photo 5842C)

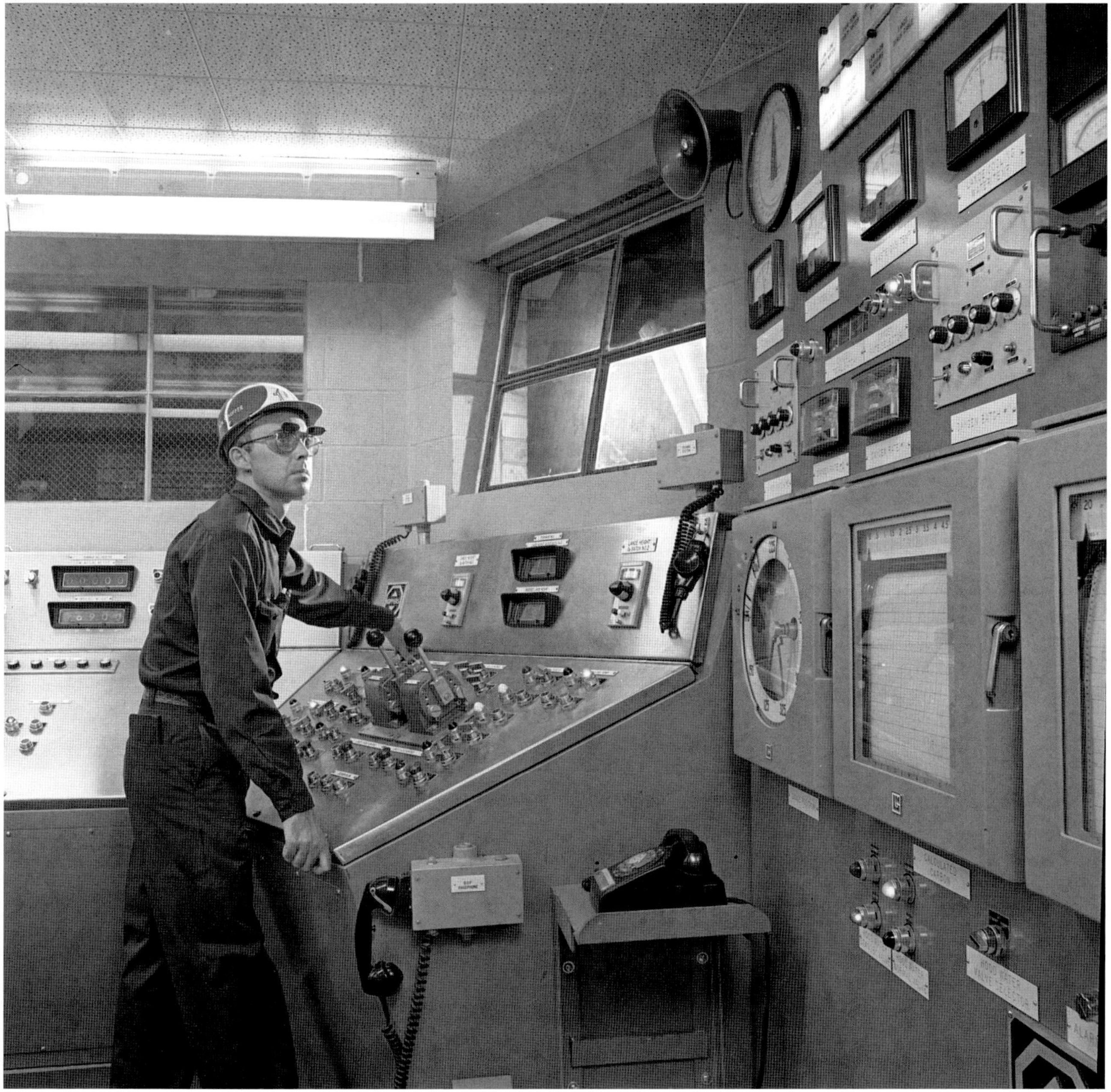

Jack Stauffer watches the gauges of his basic oxygen furnace in Cleveland from the "pulpit," which is what most control rooms in steel mills were called. Cleveland mill, 1966. (photo 6231F)

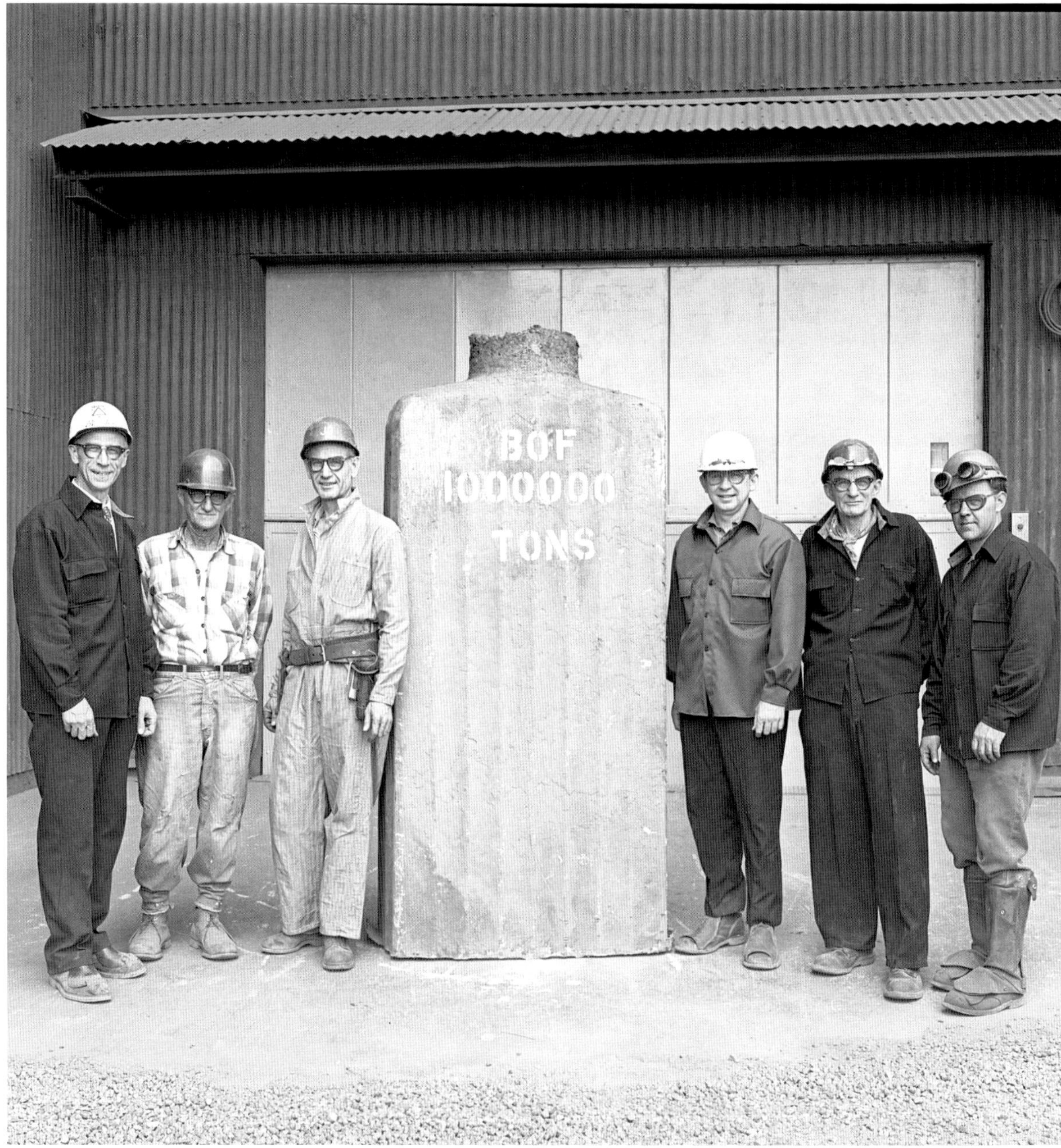

Workers from the Warren BOF (basic oxygen furnace) crew pose with an ingot on June 6, 1966, in honor of the one millionth ton of steel from the furnace. (photo 6181F)

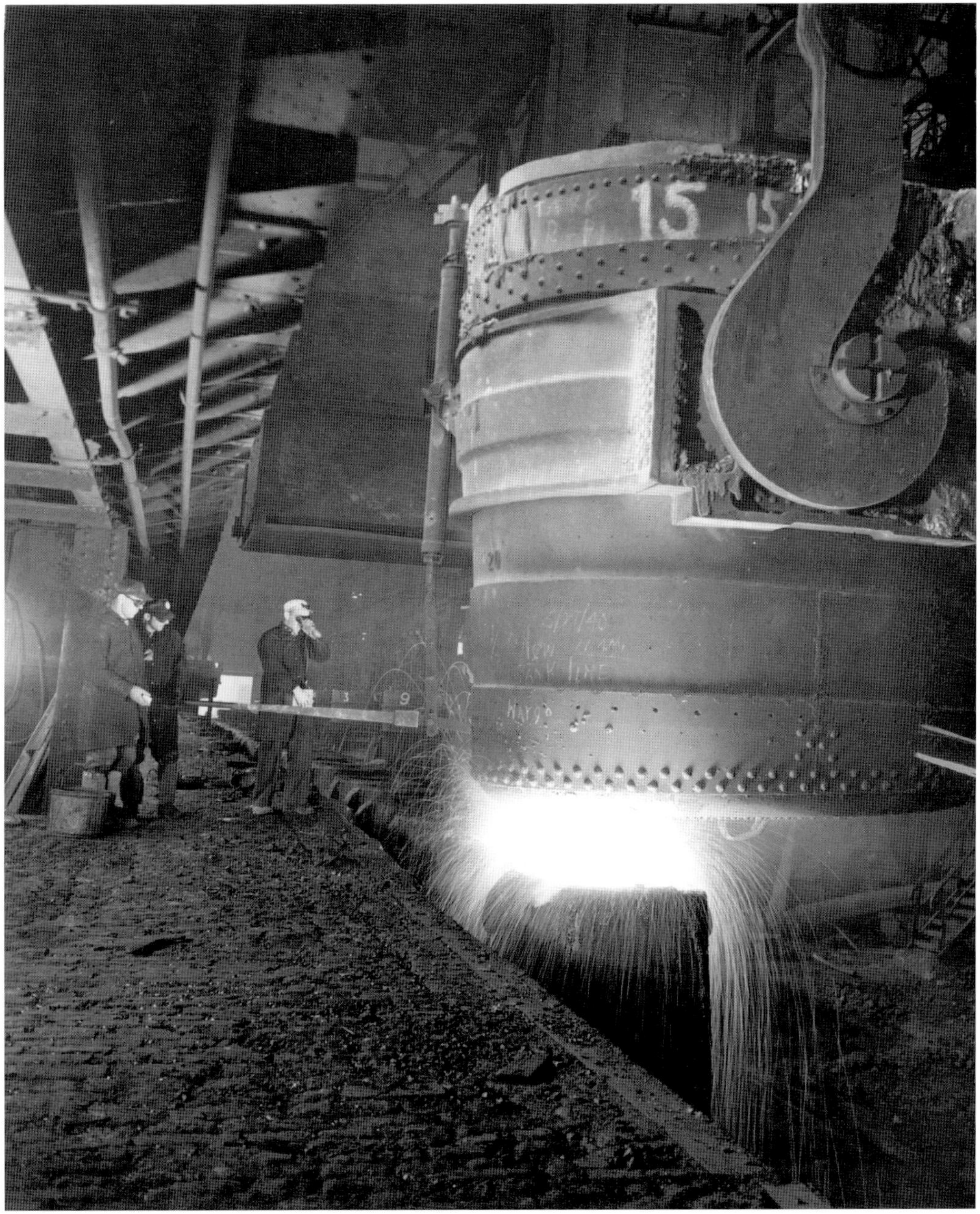

This photo, depicting the Cleveland mill in the late 1930s, shows the molten steel teeming into ingot molds.

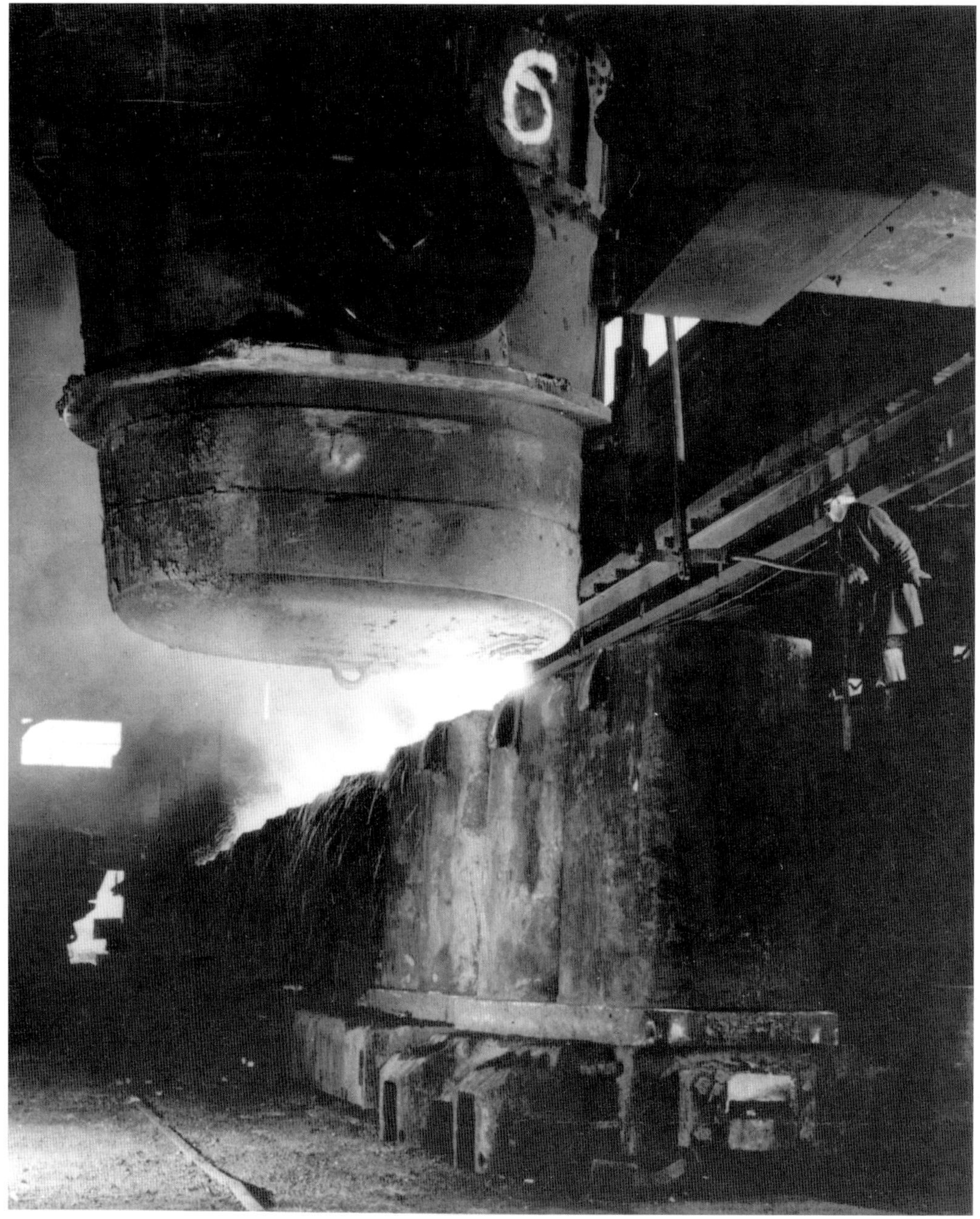

Dramatic shot of teeming into the ingot molds. Unknown mill, possibly Cleveland, ca. late 1930s.

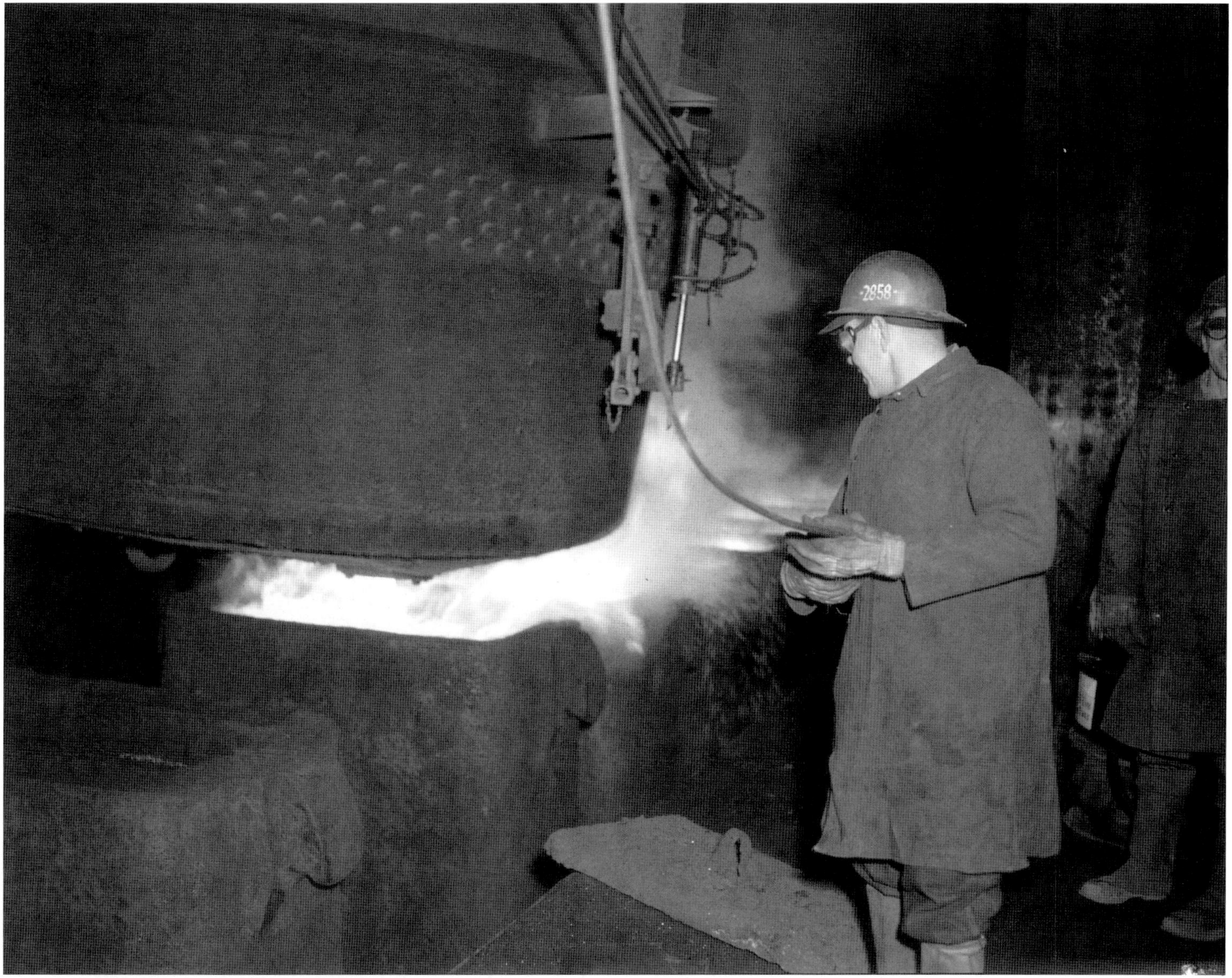

In the 1950s, Republic adopted an autopouring system for teeming. Workers, like this one, could control the process electrically, rather than manually with a lever. This photo was taken at the Cleveland mill and used as a cover for one of Republic's annual reports. (photo C1080)

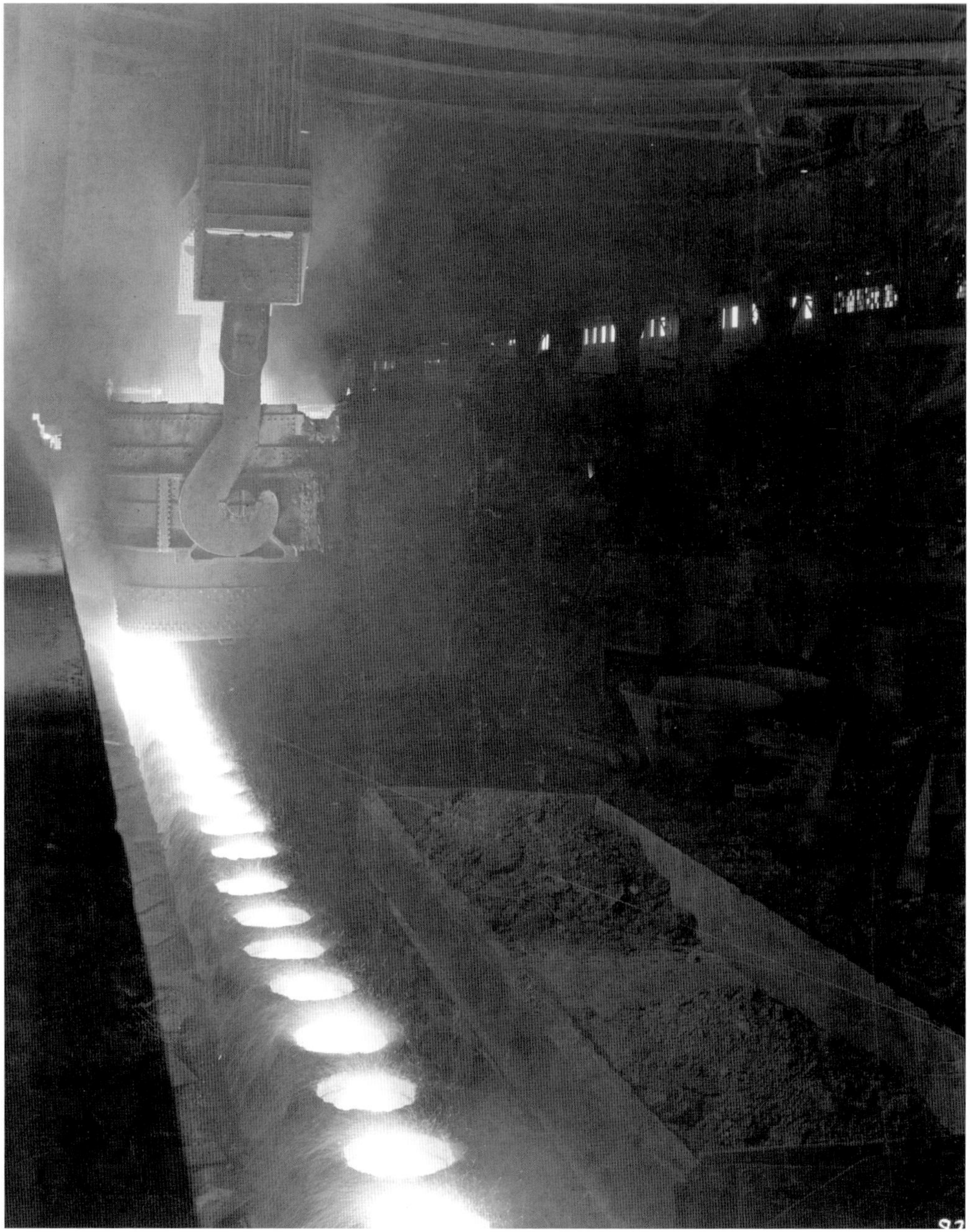

A hot ladle teems molten steel into glowing ingot molds at the Corrigan McKinney Works in Cleveland, ca. 1930s.

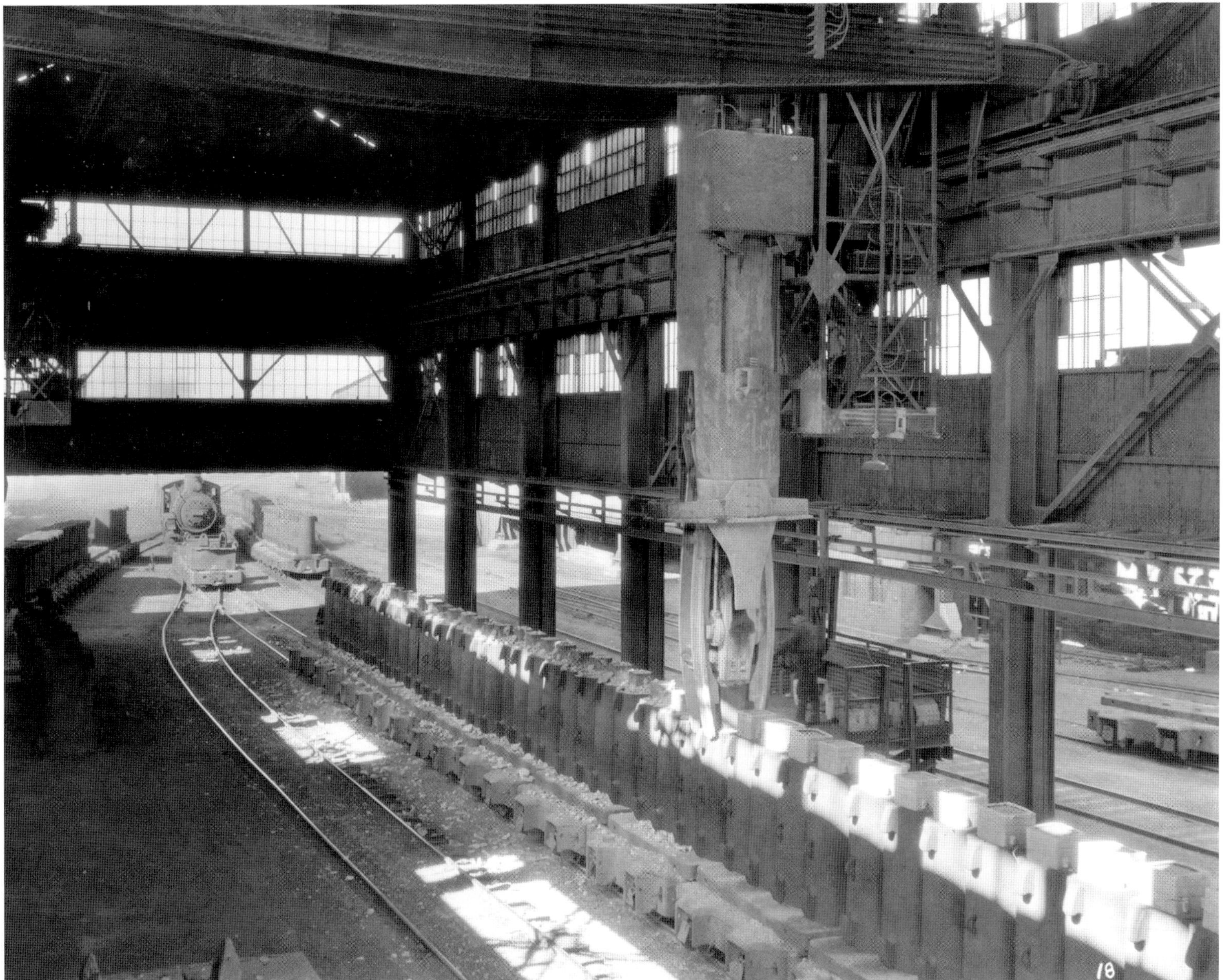

Removing hot tops from ingot molds at the Massillon plant of Republic's Central Alloy Division, ca. 1930s.

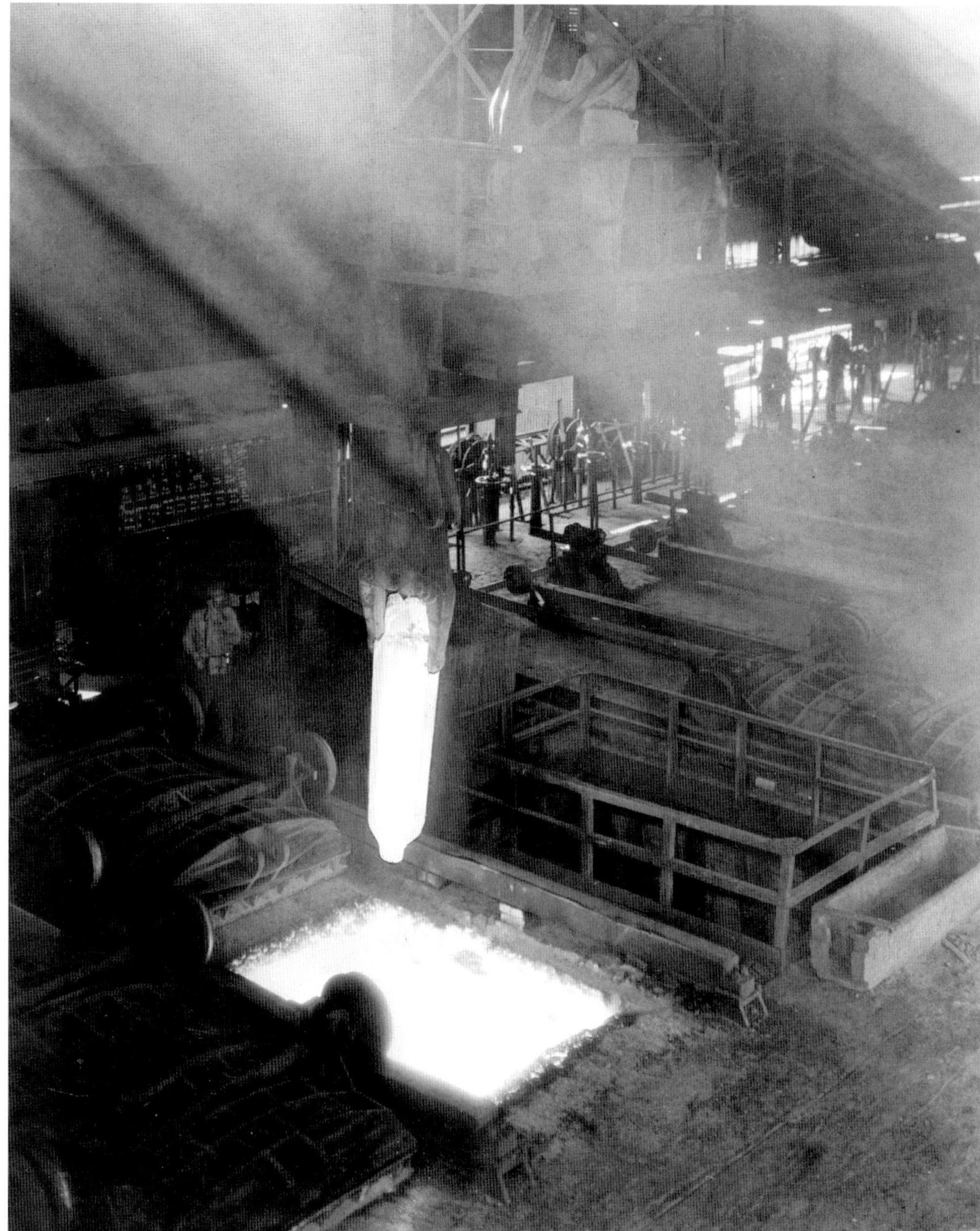

Workers placed the new steel ingots in soaking pits, which heated them to a malleable temperature. In this photo, a crane removes a hot ingot from the soaking pit at Republic's Massillon mill, ca. 1930s.

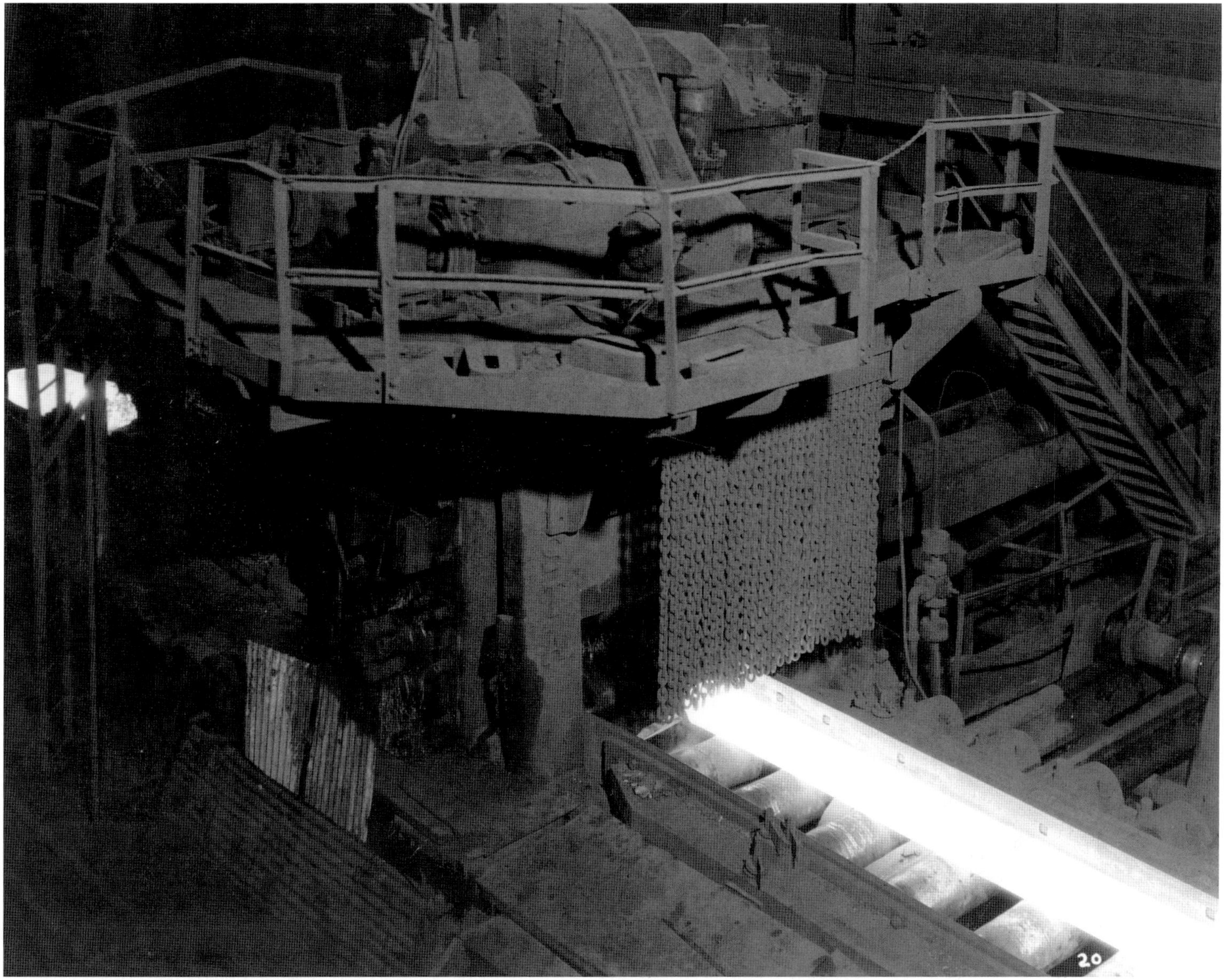

After emerging from the soaking pit, the hot ingot was run through the blooming mill, which lengthened and flattened the steel into a bloom. In this photo, a bloom is leaving the mill and heading for the shears to be trimmed. At the left of the photo, an ingot just enters the blooming mill. Republic's Massillon mill, ca. 1930s.

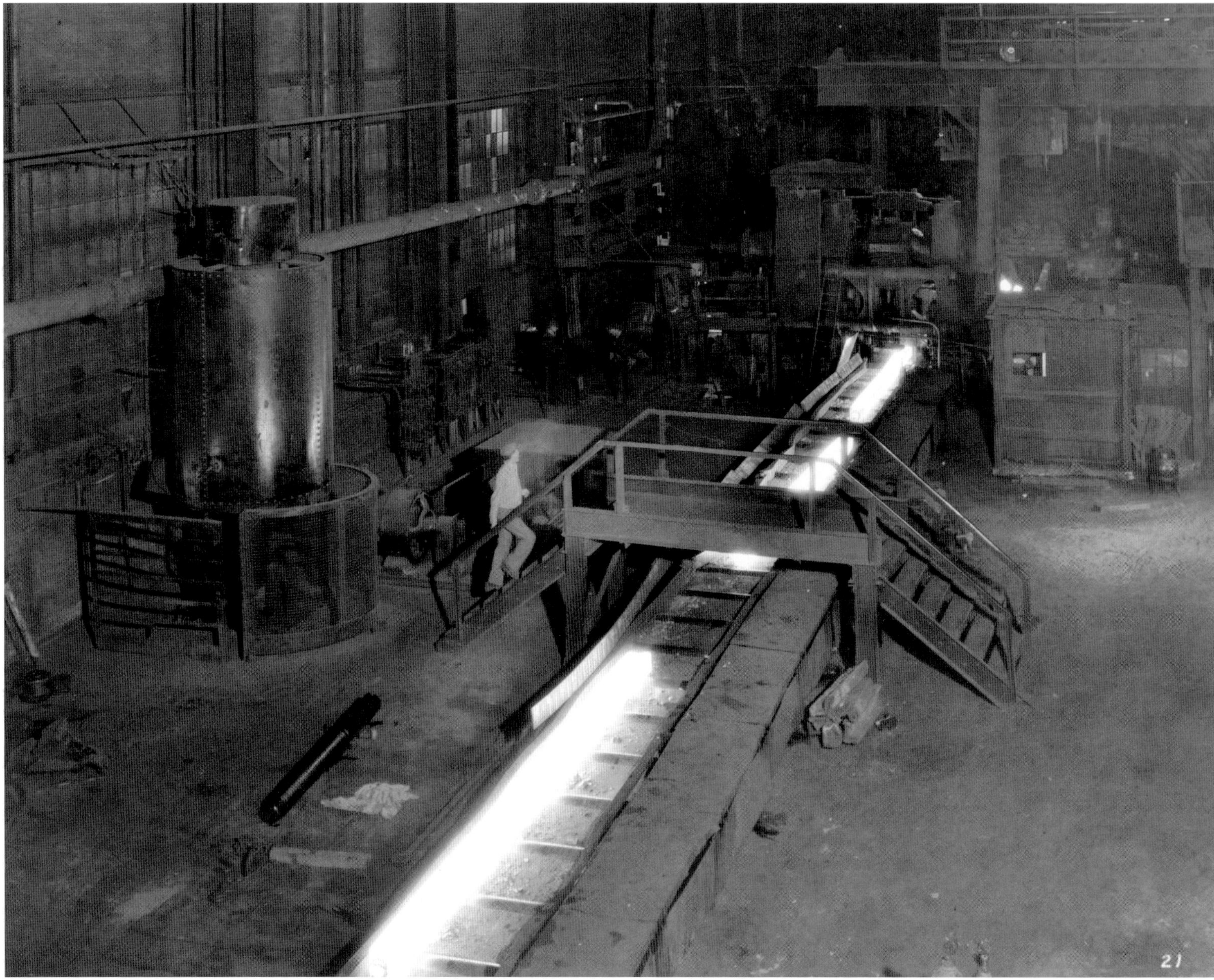

This photo shows steel blooms approaching shears that will crop the ends. Republic's Massillon mill, ca. 1930s.

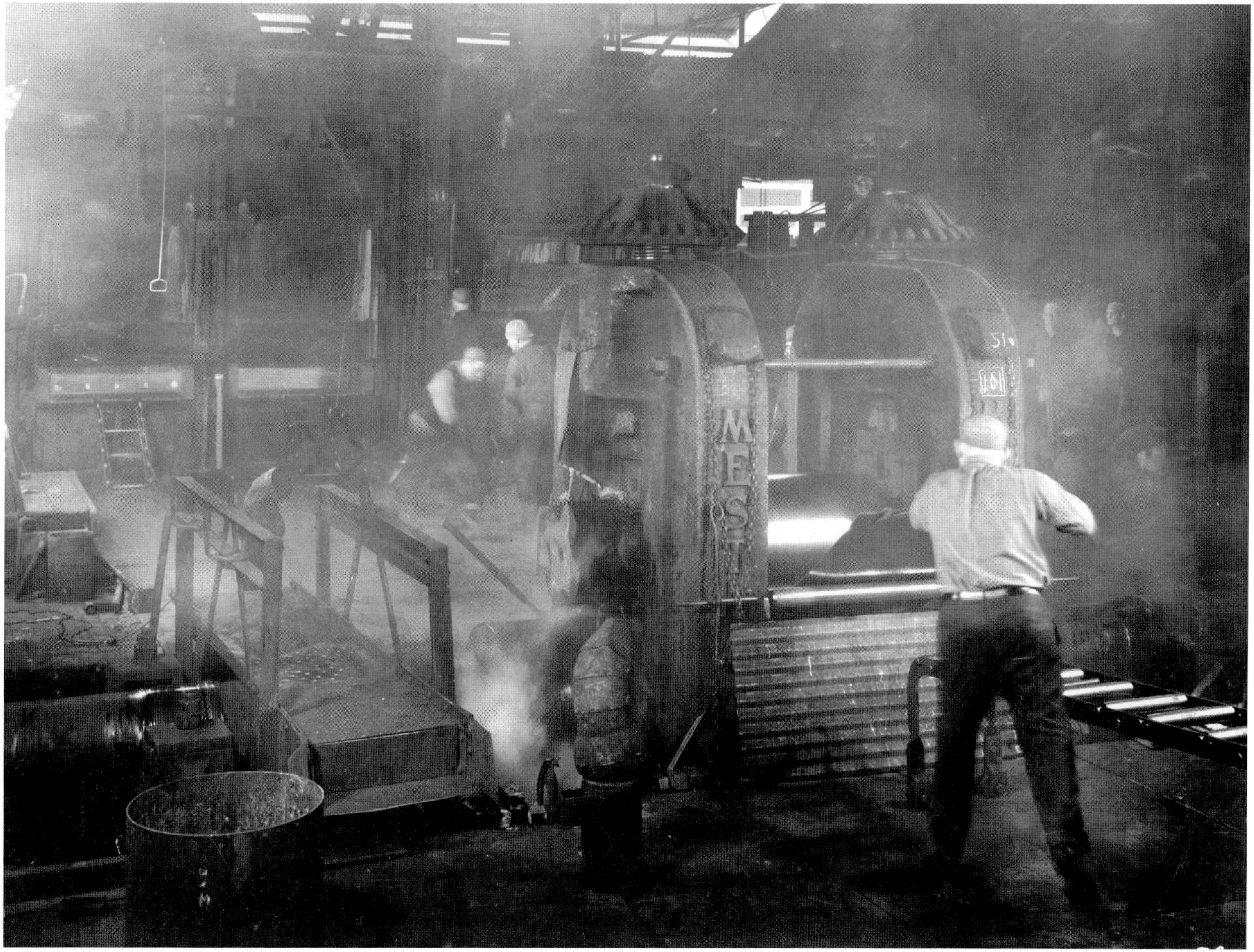

An old-style hot sheet mill rolls silicon sheets. The worker on the right skillfully uses tongs to catch the hot sheet of steel and pass it back through the mill to elongate it. Republic's Warren, Ohio, plant, ca. 1930s.

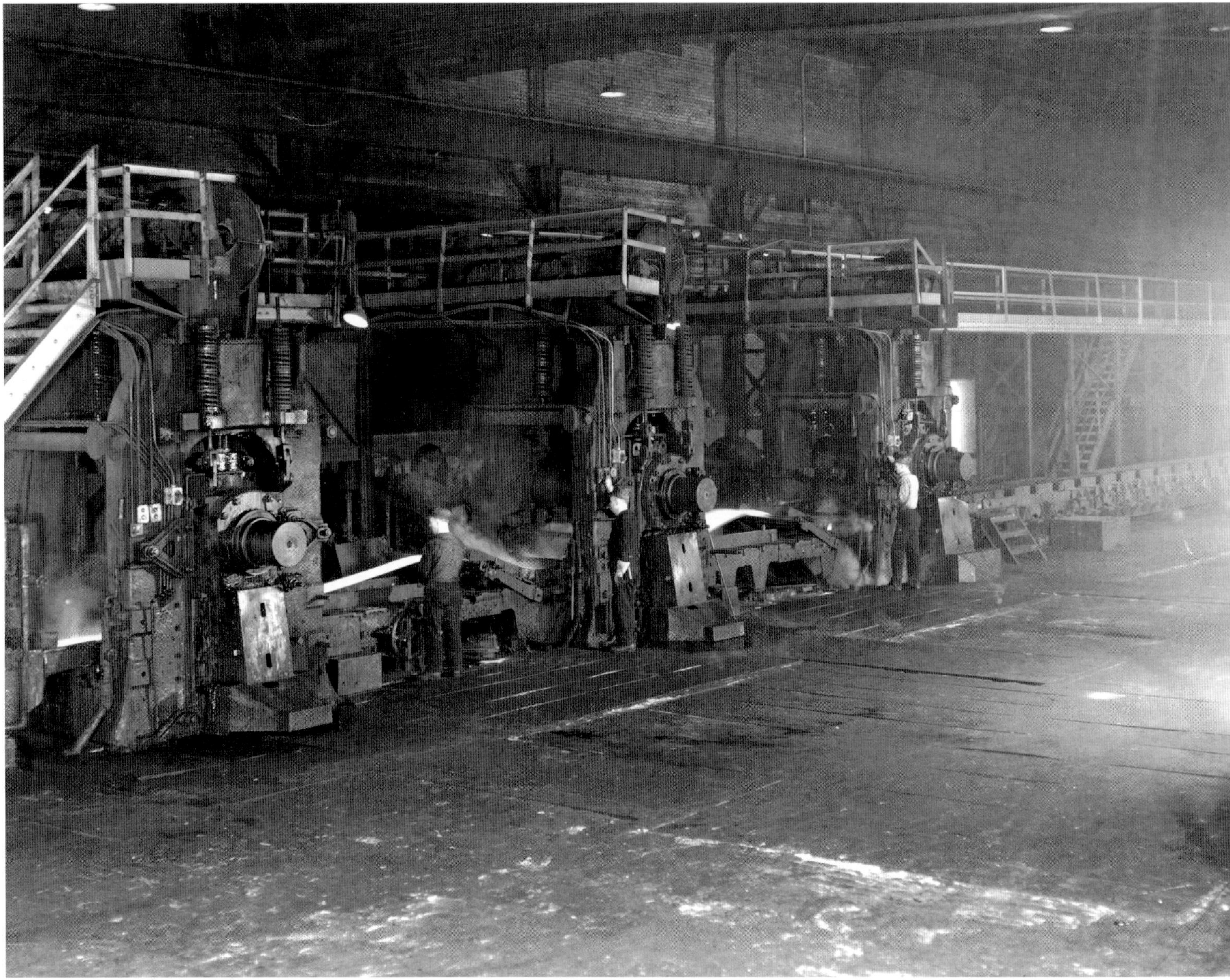

In this automated hot strip mill successive roll stands squeeze the hot blooms thinner and thinner as they move along the rolling mill in Republic's Warren Works, ca. 1930s.

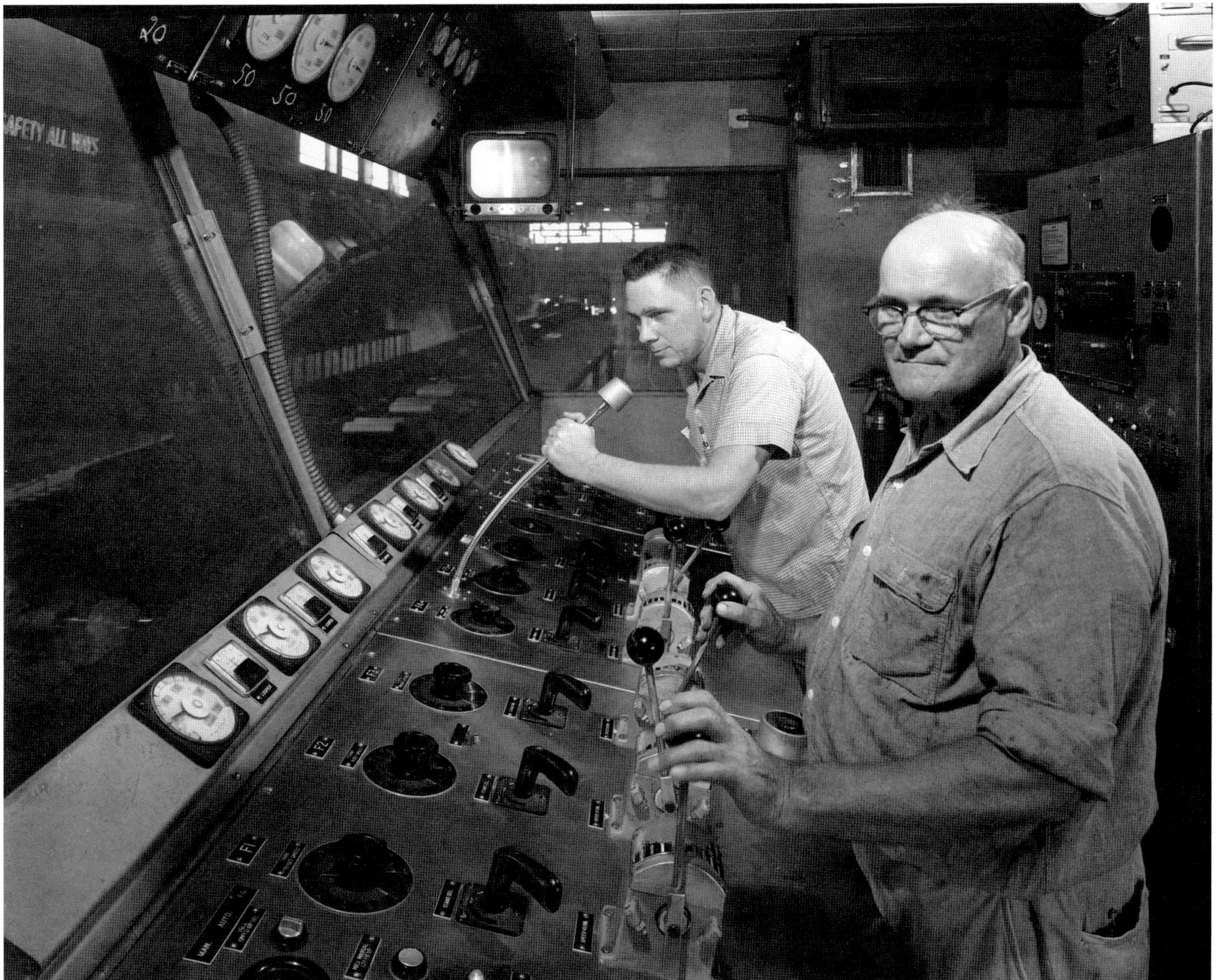

The Warren 56-inch hot strip mill pulpit on June 15, 1965, with Elmer Boonan, finisher, in the foreground and Paul Moser, assistant roller, in the background. Today, a hot mill pulpit looks more like the control deck of a space shuttle, as it is heavily computerized, with workers observing the process rather than physically controlling it. (photo 5224F)

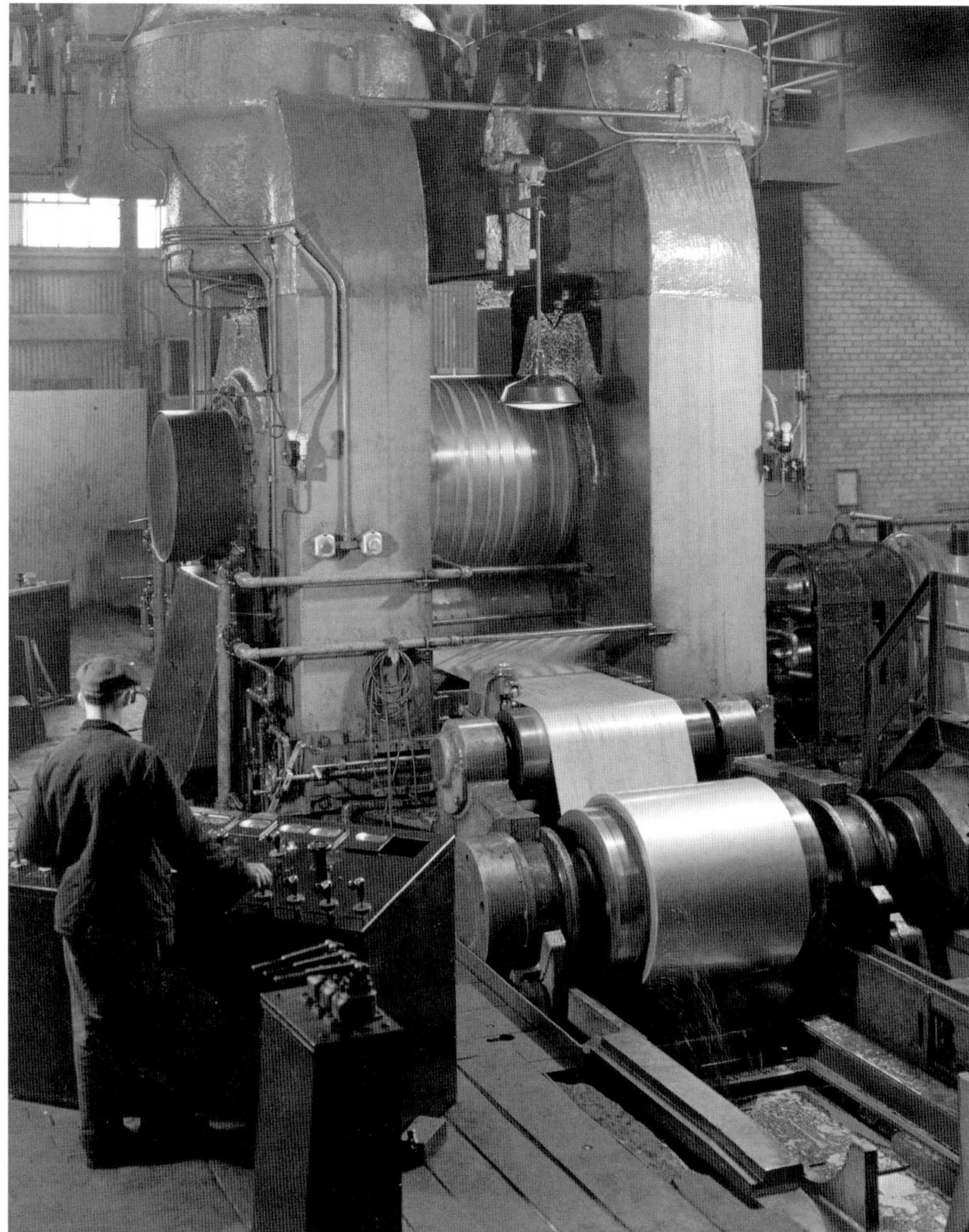

After the hot mill flattened the steel, it was often rerolled in a cold mill, such as this one, to further reduce its thickness and provide a better finish. Republic's Warren mill, ca. 1930s.

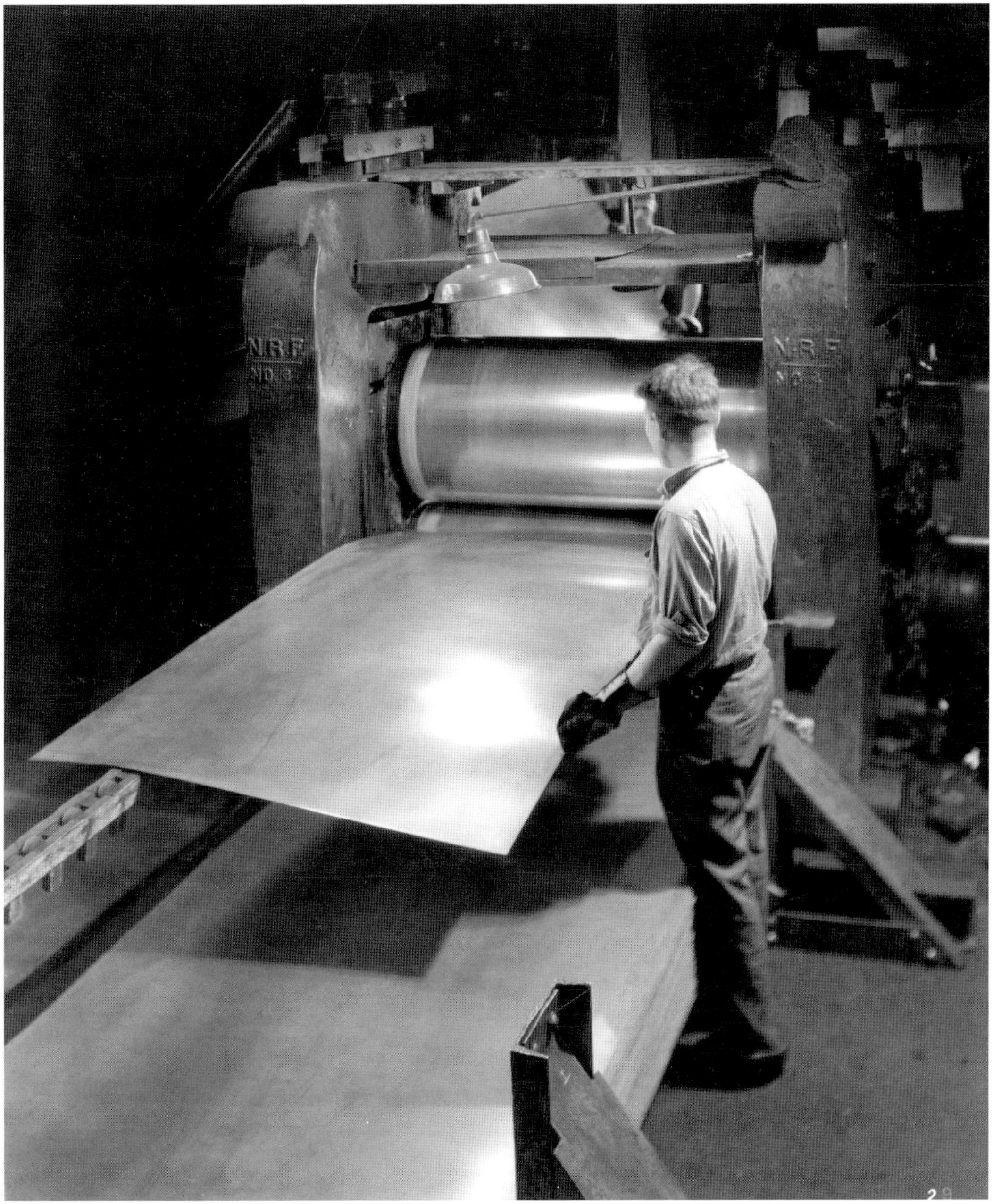

A sheet of steel is cold-rolled at Republic's Massillon mill.

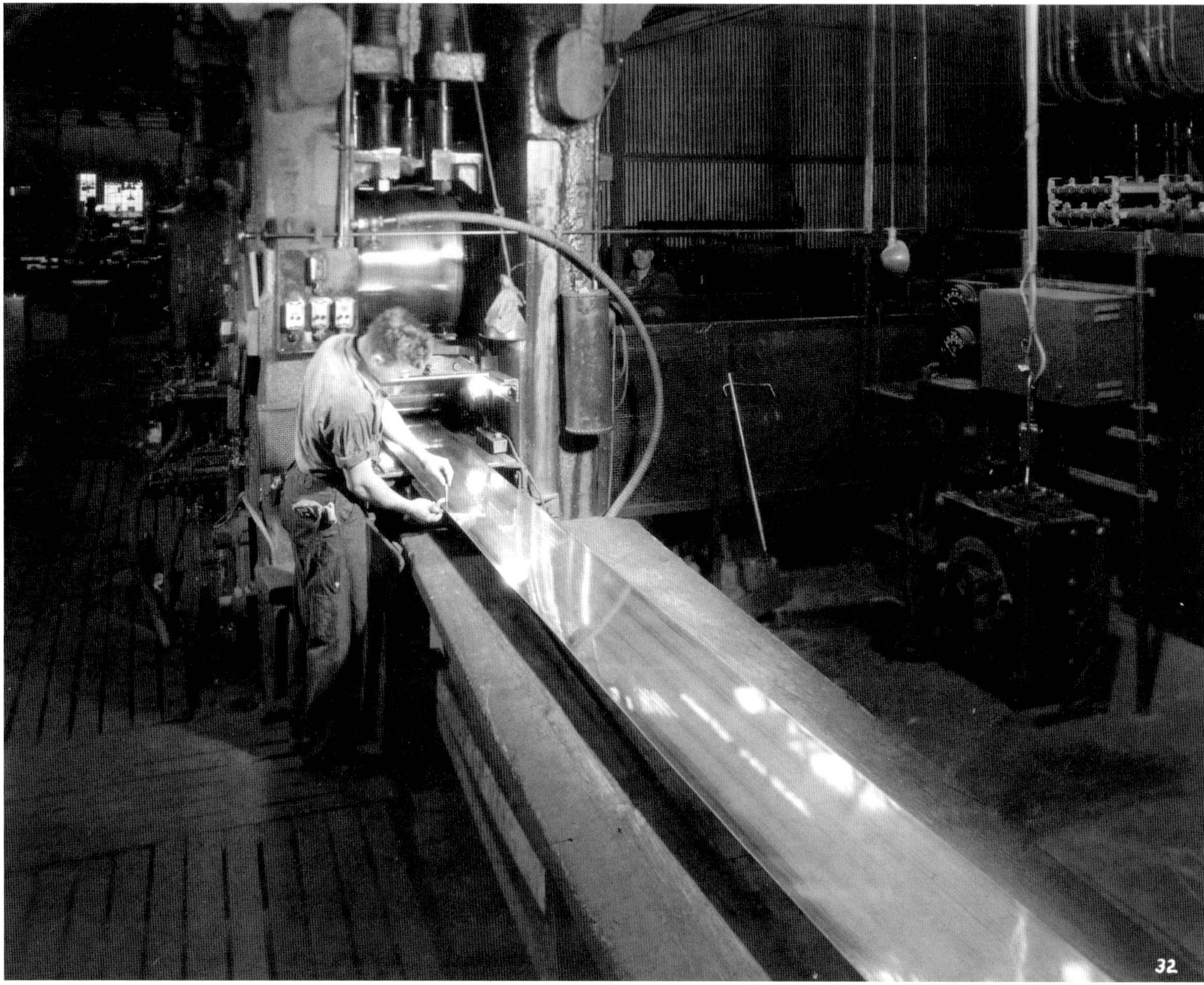

This worker uses a caliper to check the thickness of Enduro stainless steel that is being cold-rolled. Today computers measure the steel. Massillon mill, ca. 1930s.

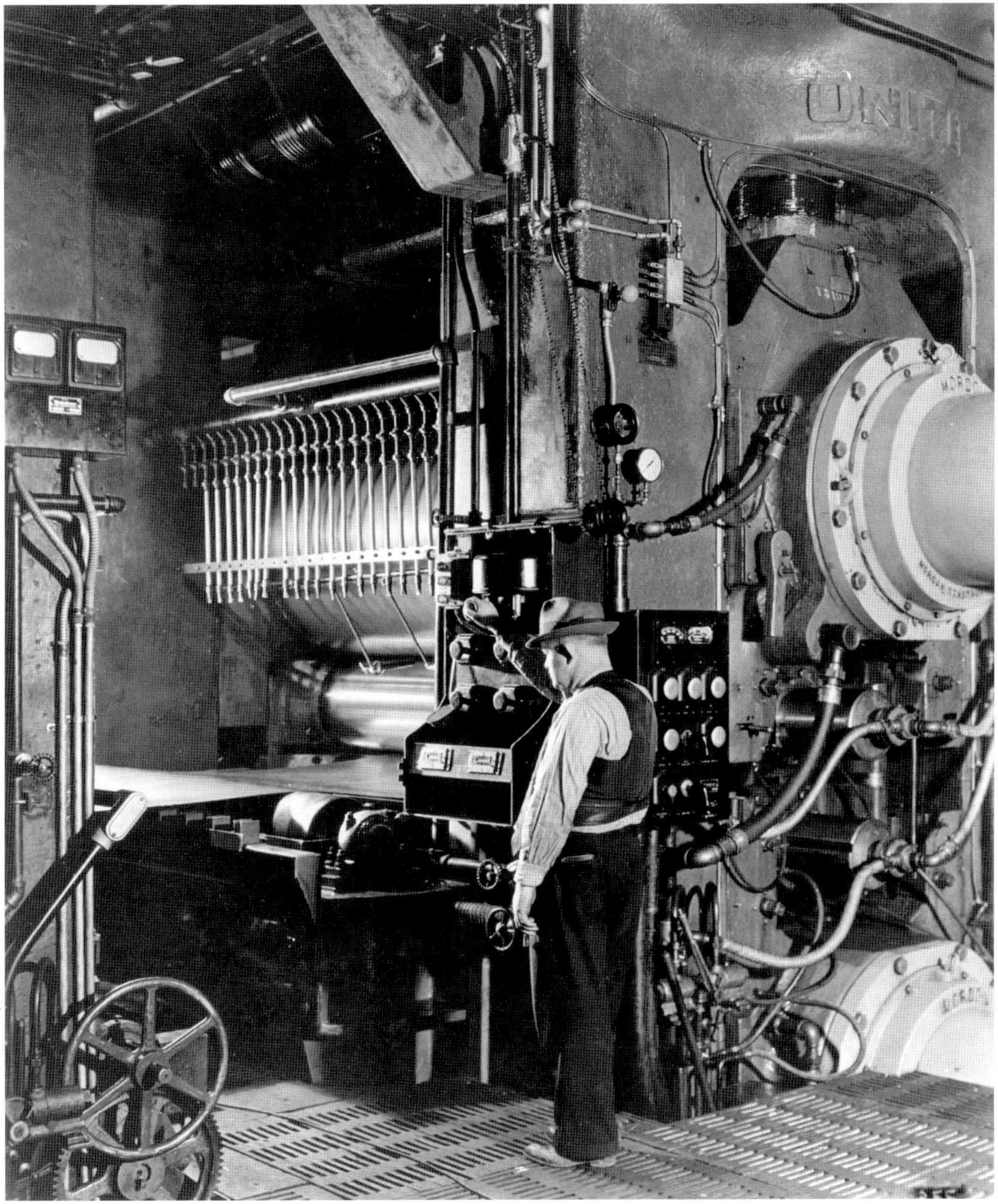

Strip steel being processed in the cold reduction mill at Republic Steel's 98-inch continuous strip mill in Cleveland, which was built in 1937, shortly before this photo was taken.

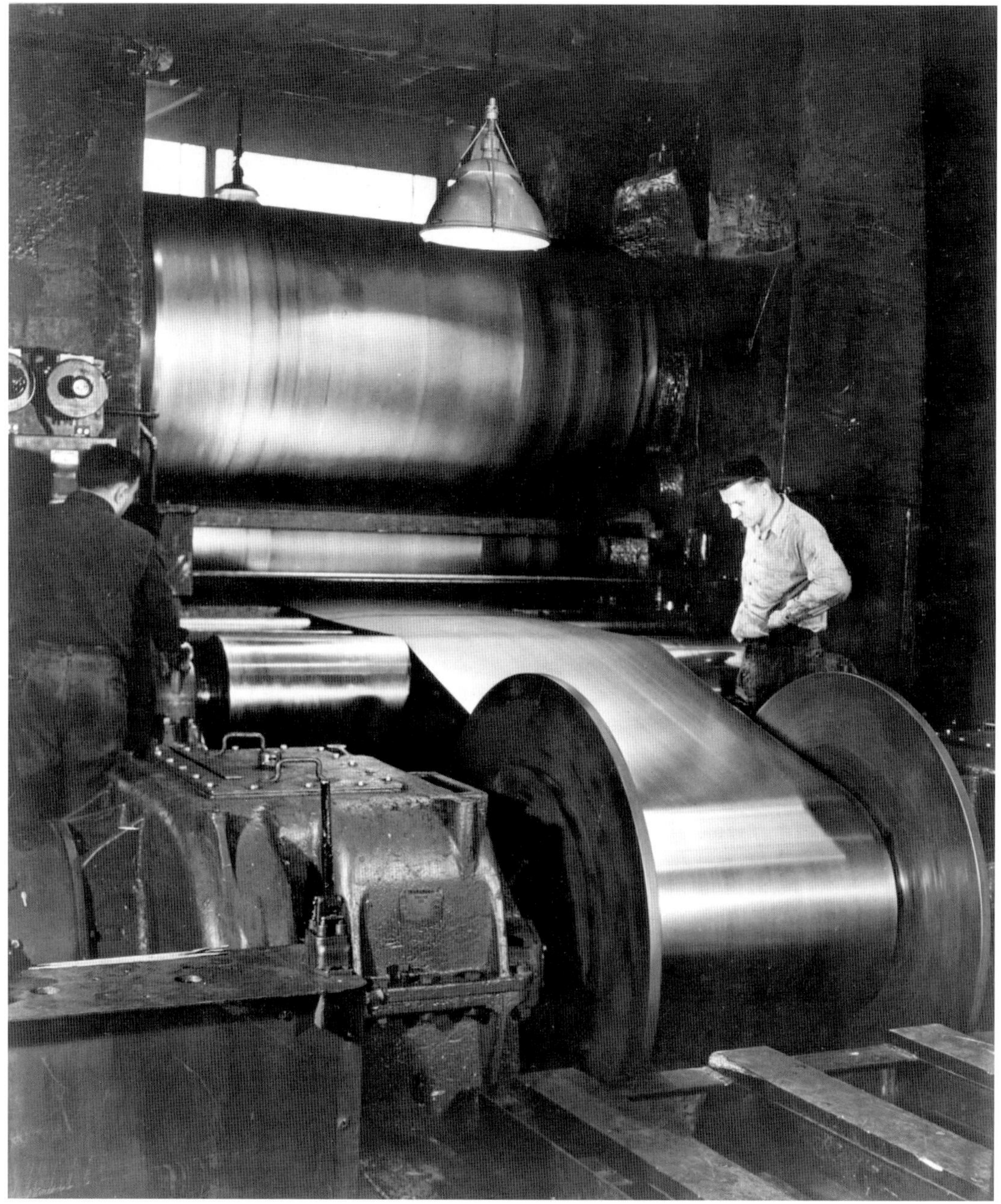

Recoiling of strip after rolling in the 98-inch mill at Republic Steel's continuous strip mill in Cleveland. This mill is a combination unit that can handle either coils or sheets, ca. 1937.

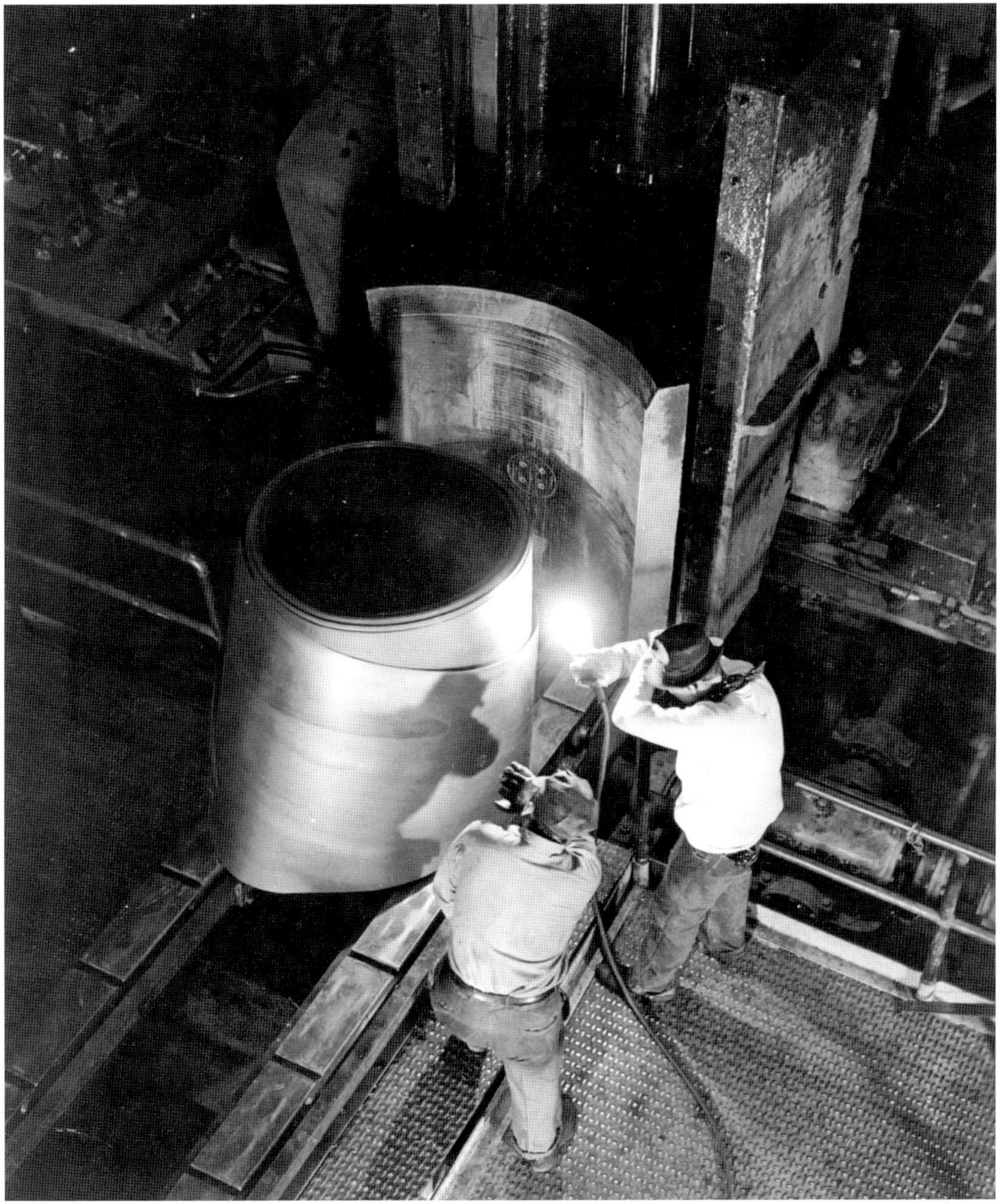

Strip steel was coiled at speeds up to 2500 feet per minute in coilers at Republic Steel's 98-inch continuous hot mill in Cleveland. These workers closely inspect the coil on a conveyor, shielding themselves from the intense heat, which is still hundreds of degrees, ca. 1937.

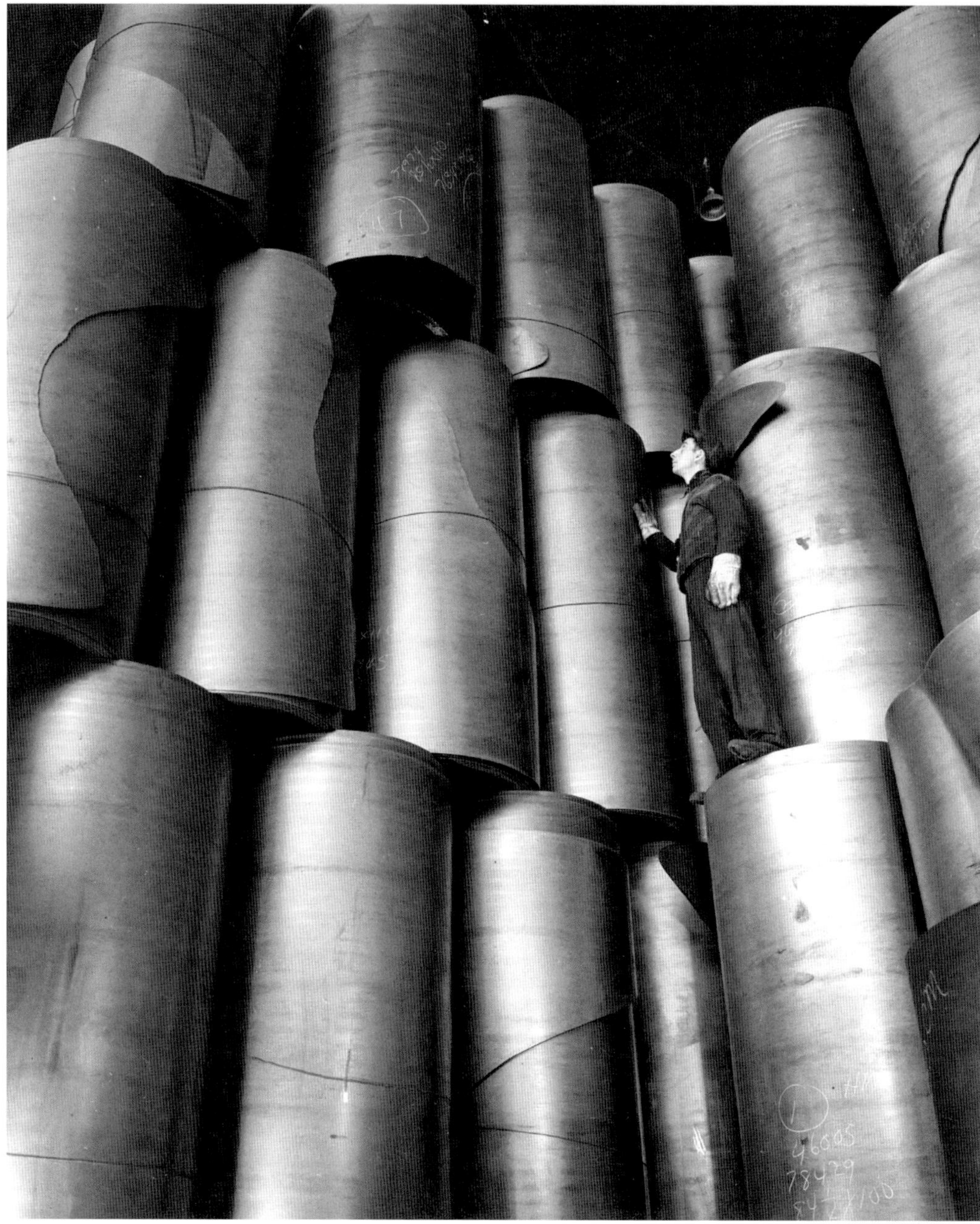

A great publicity shot, losing the man among the giant coils of steel, ca. 1940. Photo by C. W. Ackerman Studio, Cleveland.

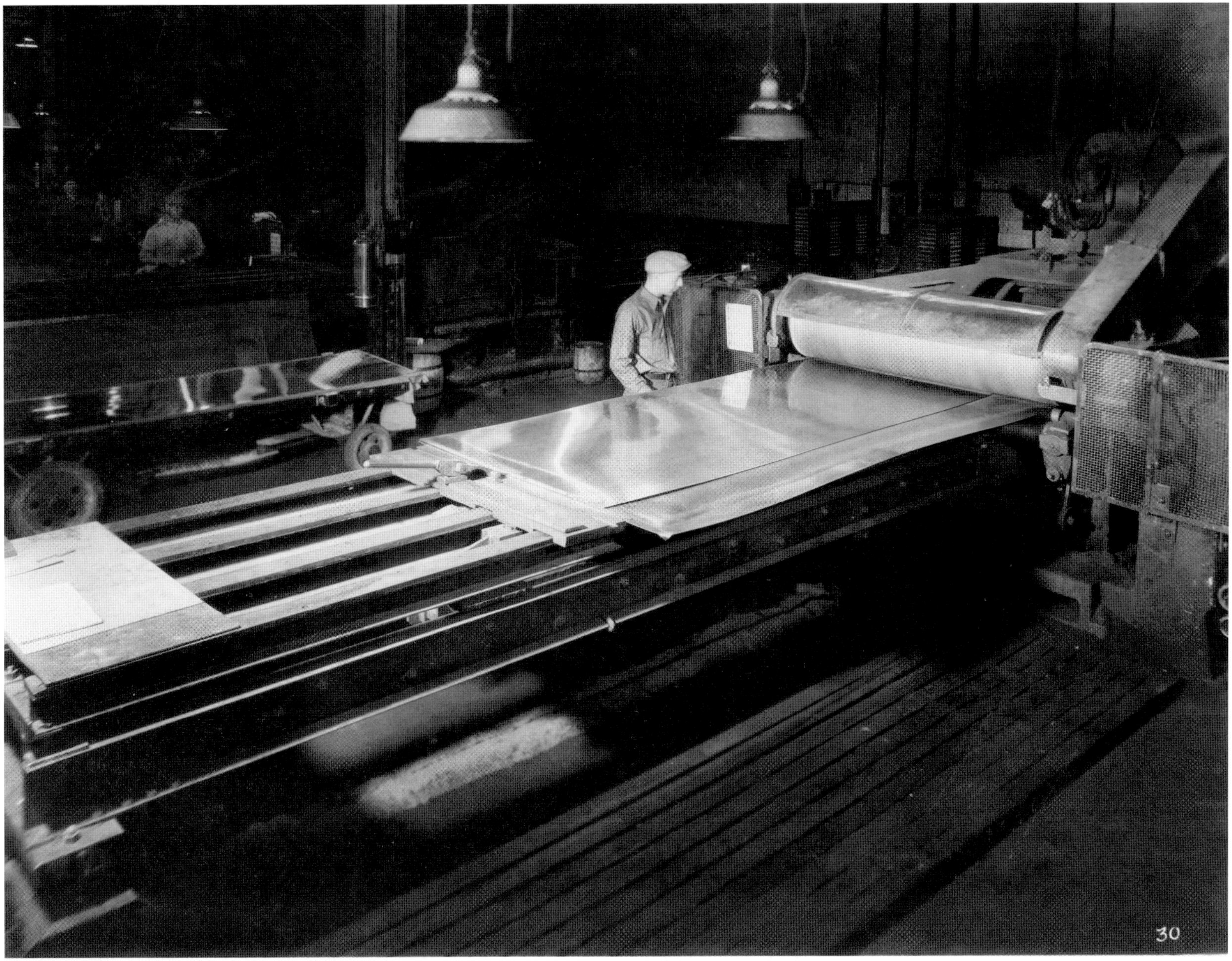

A worker polishes a sheet of Enduro stainless steel at the Massillon mill.

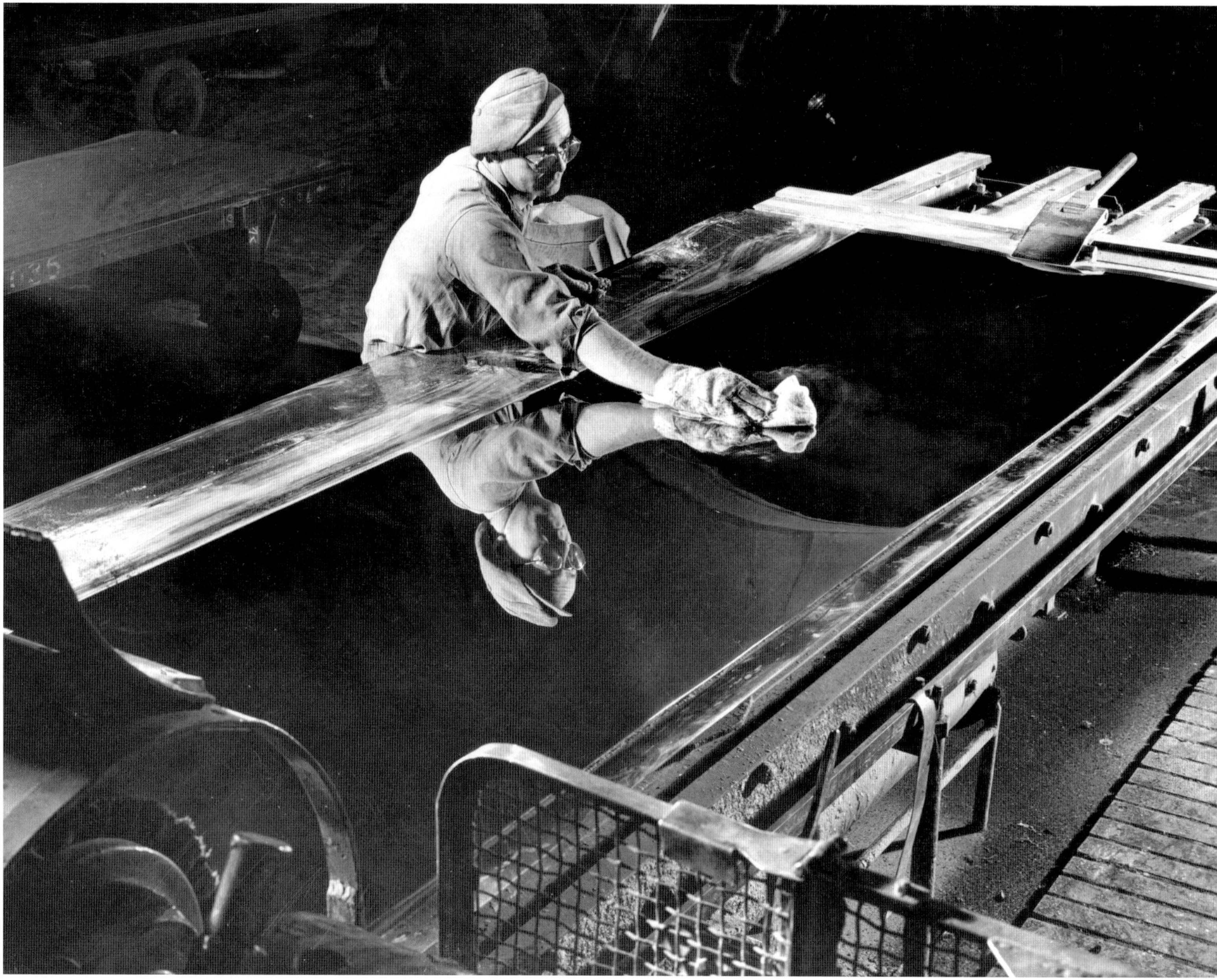

A worker polishes stainless steel, either in the Canton or Massillon mill, in 1941. Photo by C. W. Ackerman Studio, Cleveland.

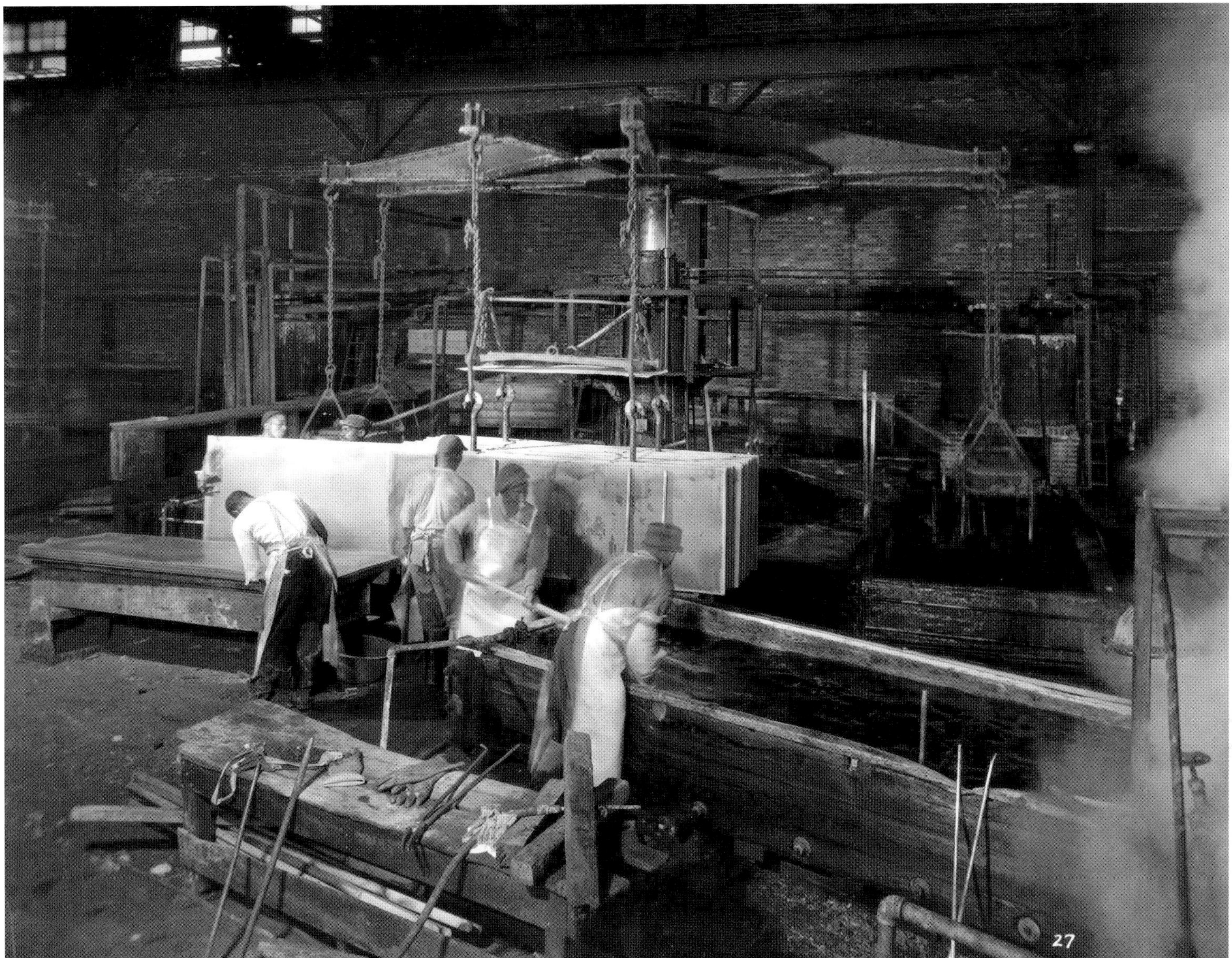

This photo from the Massillon mill shows pickling, when steel sheets were dipped in an acid bath to cleanse the surface. Notice that in this unpleasant environment, the entire workforce is African American, and the safety equipment consists primarily of aprons and thick gloves, most of which are going unused, ca. late 1930s.

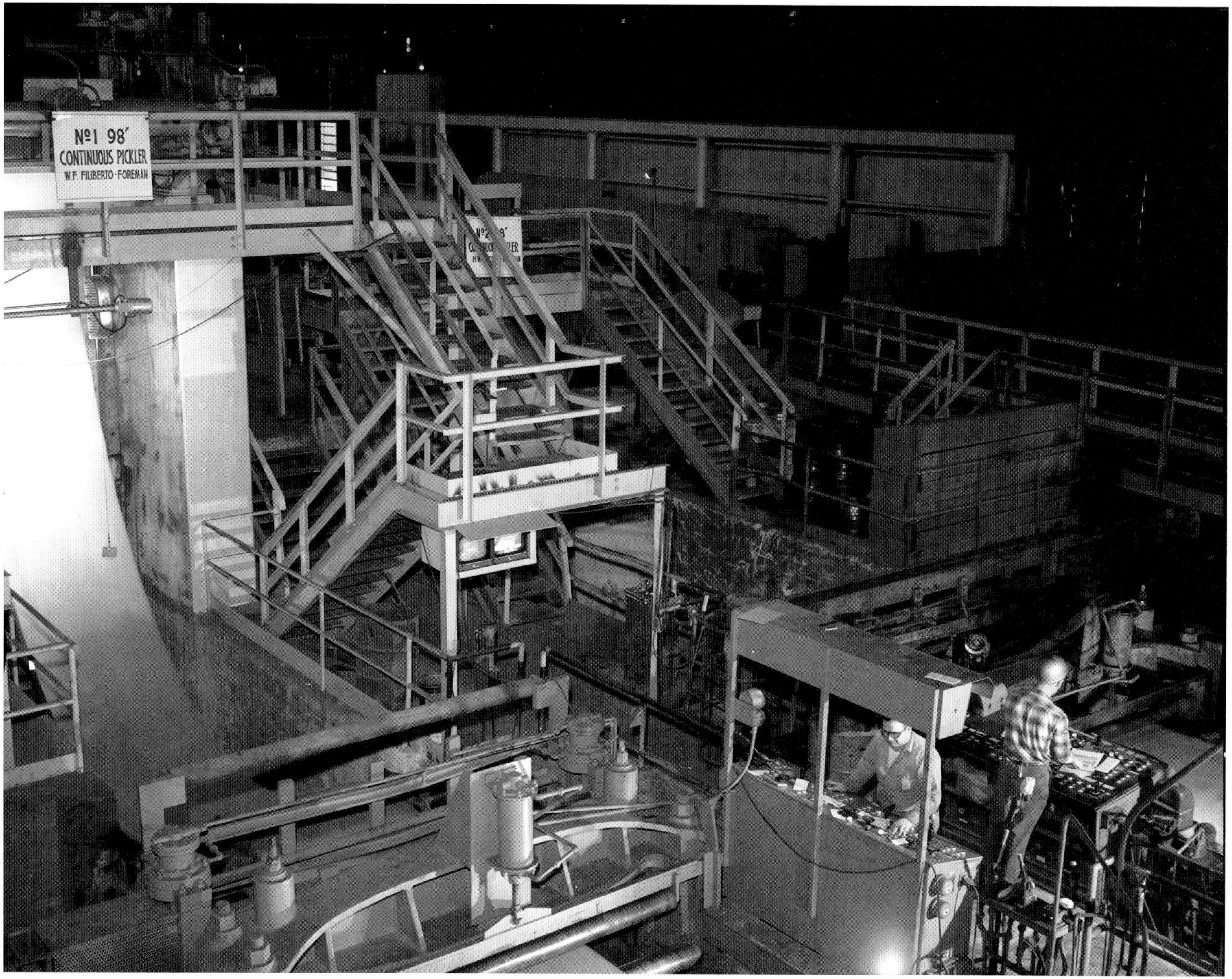

Pickling in 1966 at the Cleveland mill's 98-inch continuous pickler, which is a lot cleaner and less dangerous than pickling in the previous photo. (photo 6145B)

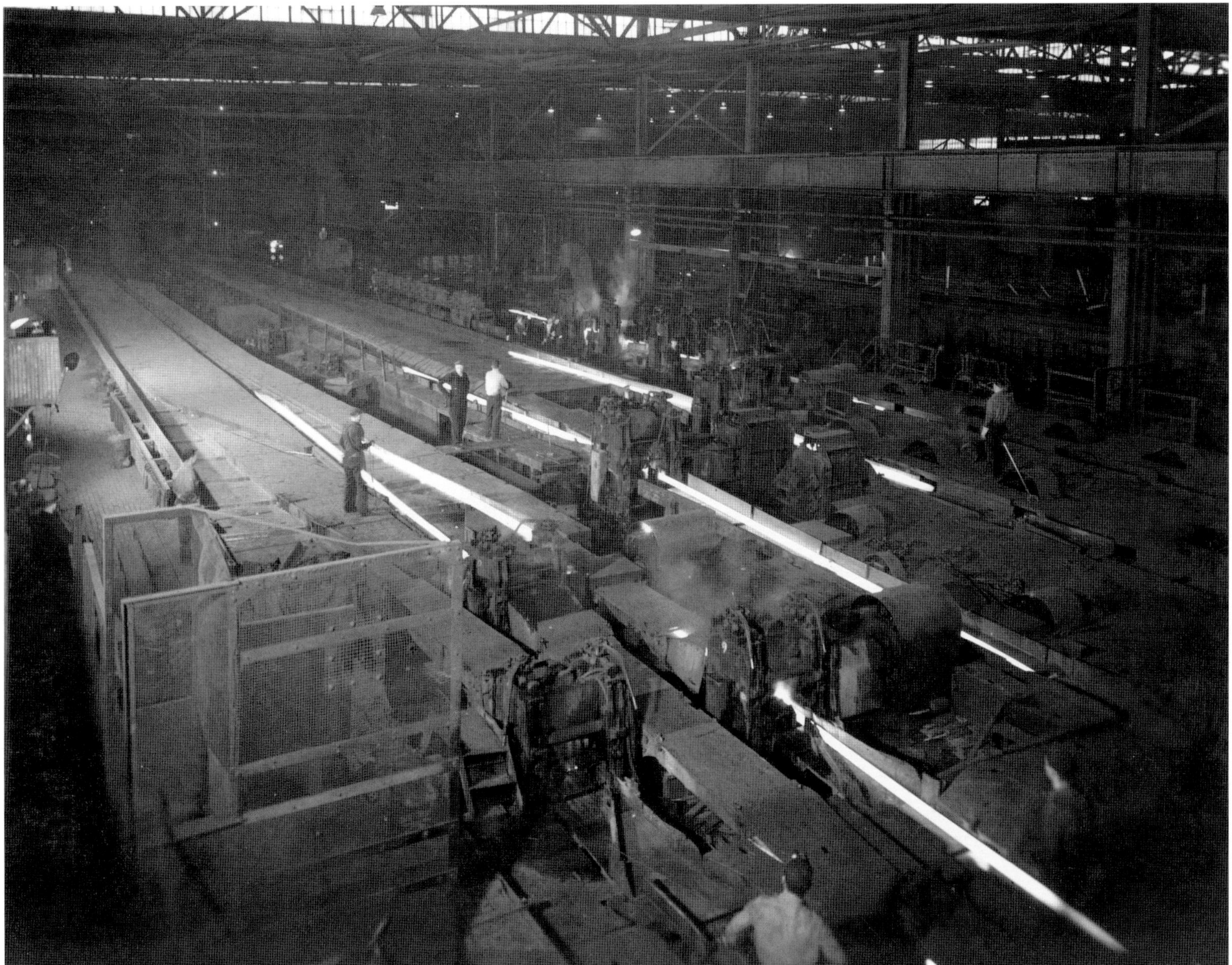

In addition to sheet steel, Republic's mills made steel bars that were used in the automobile industry. This photo shows the runout tables, where the bars would cool, at the end of the bar mill at the Cleveland mill, late 1930s.

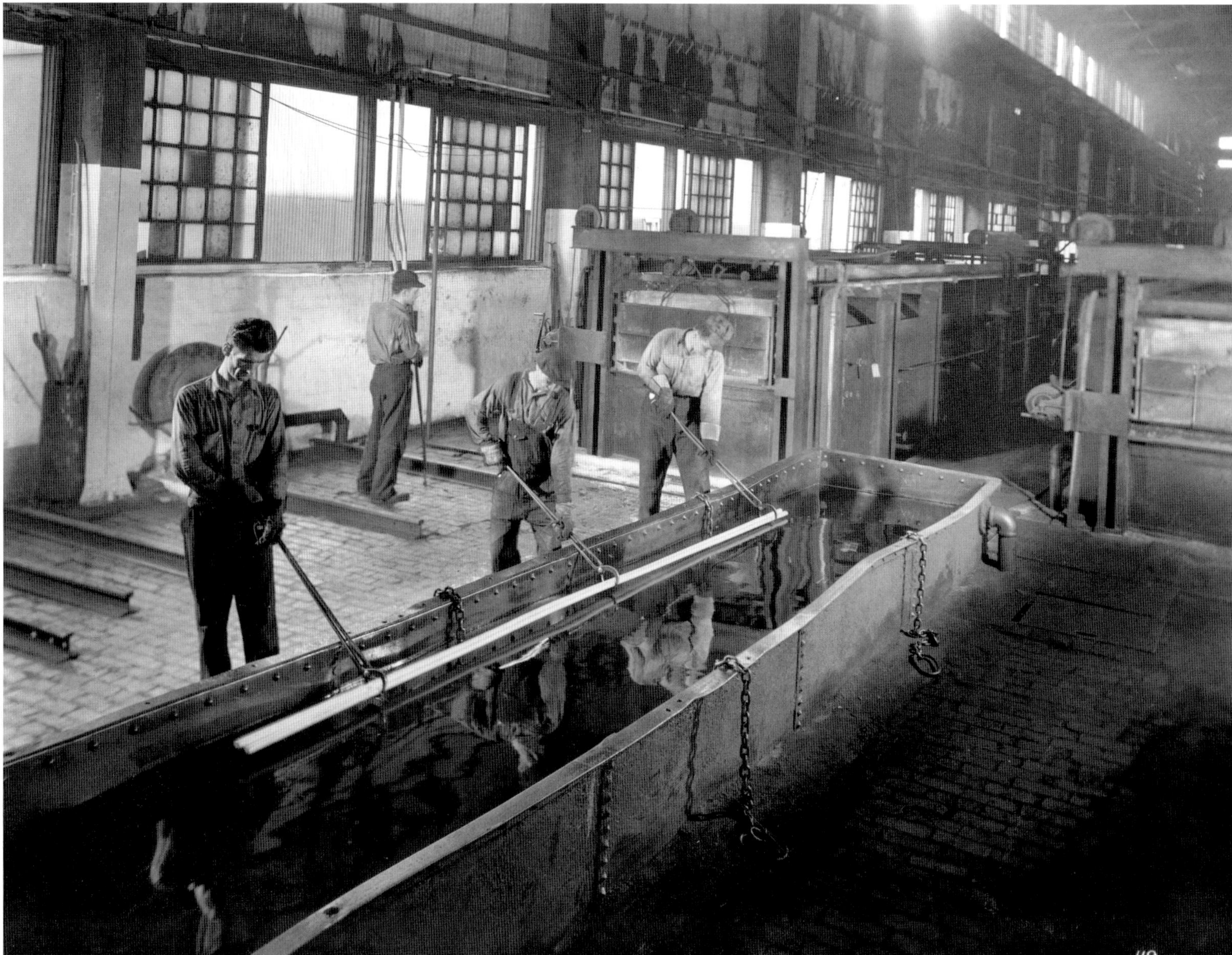

In this 1930s photo of the Canton mill, heat-treating furnaces in the background were used for annealing, quenching, drawing, and normalizing small sections of steel bars. In the foreground workers quench alloy bars still hot from the furnaces.

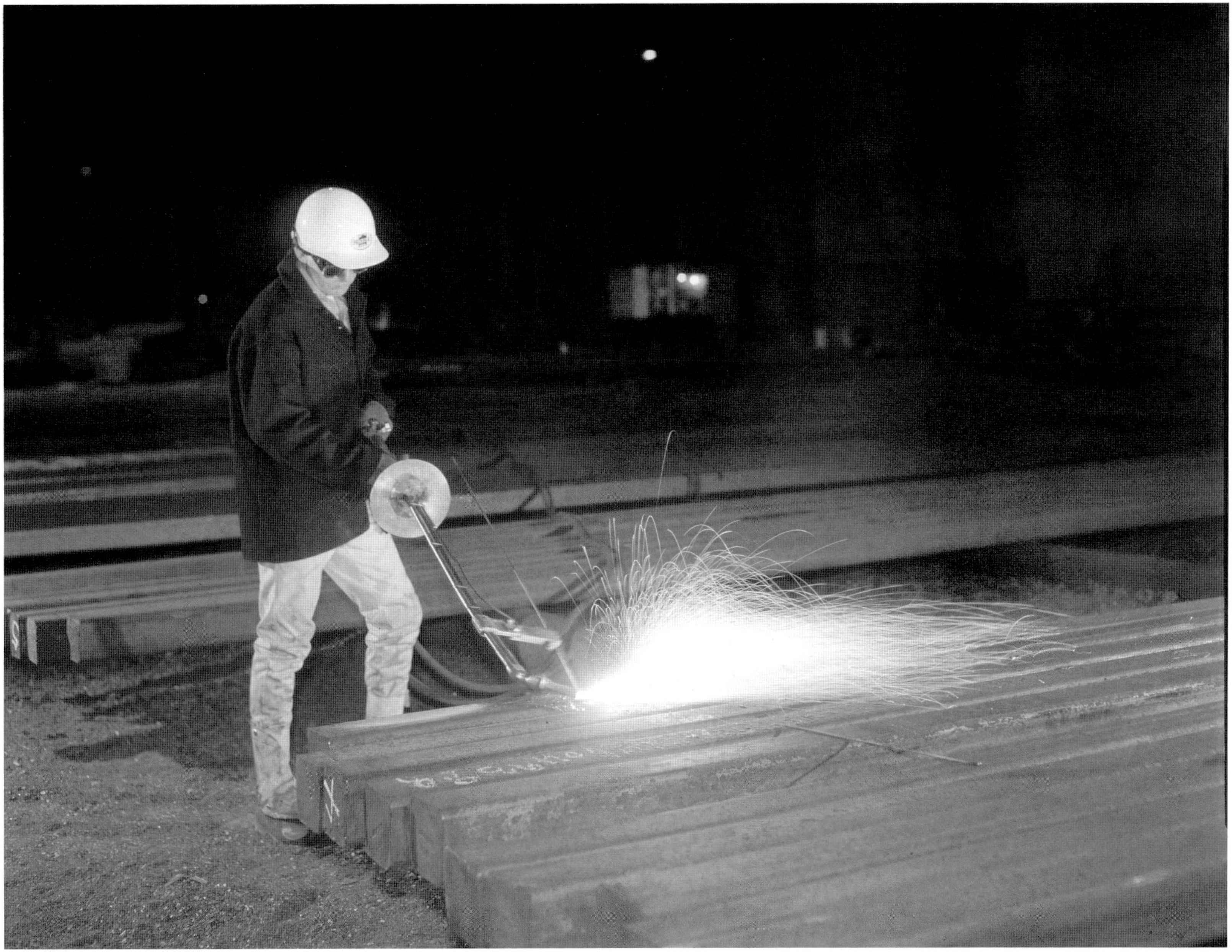

This worker uses a torch to remove surface imperfections from steel in the Cleveland bar mill, ca. 1960s. (photo C1269)

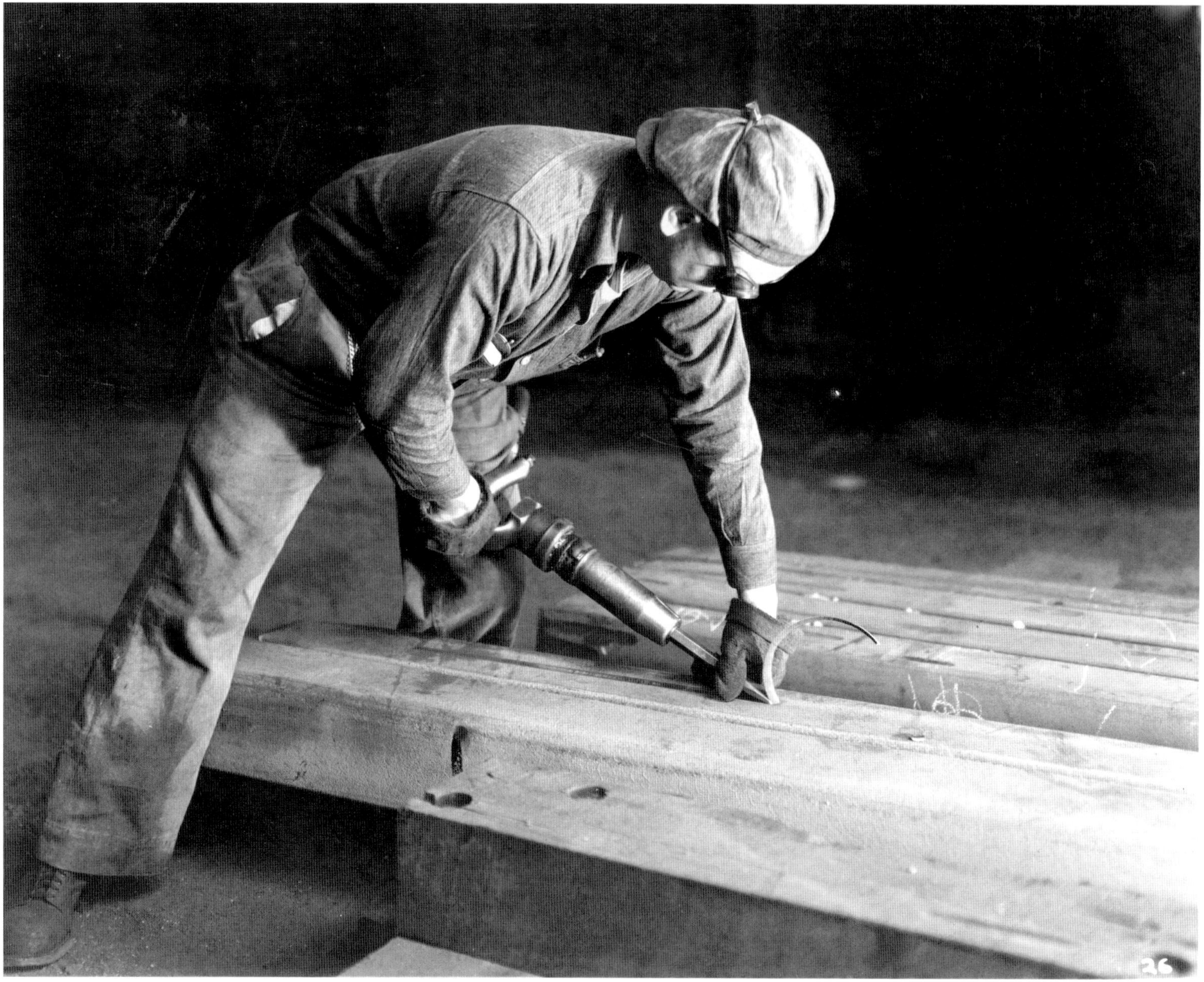

This worker uses a pneumatic hammer to remove surface defects from an alloy steel billet in the Massillon mill in the late 1930s.

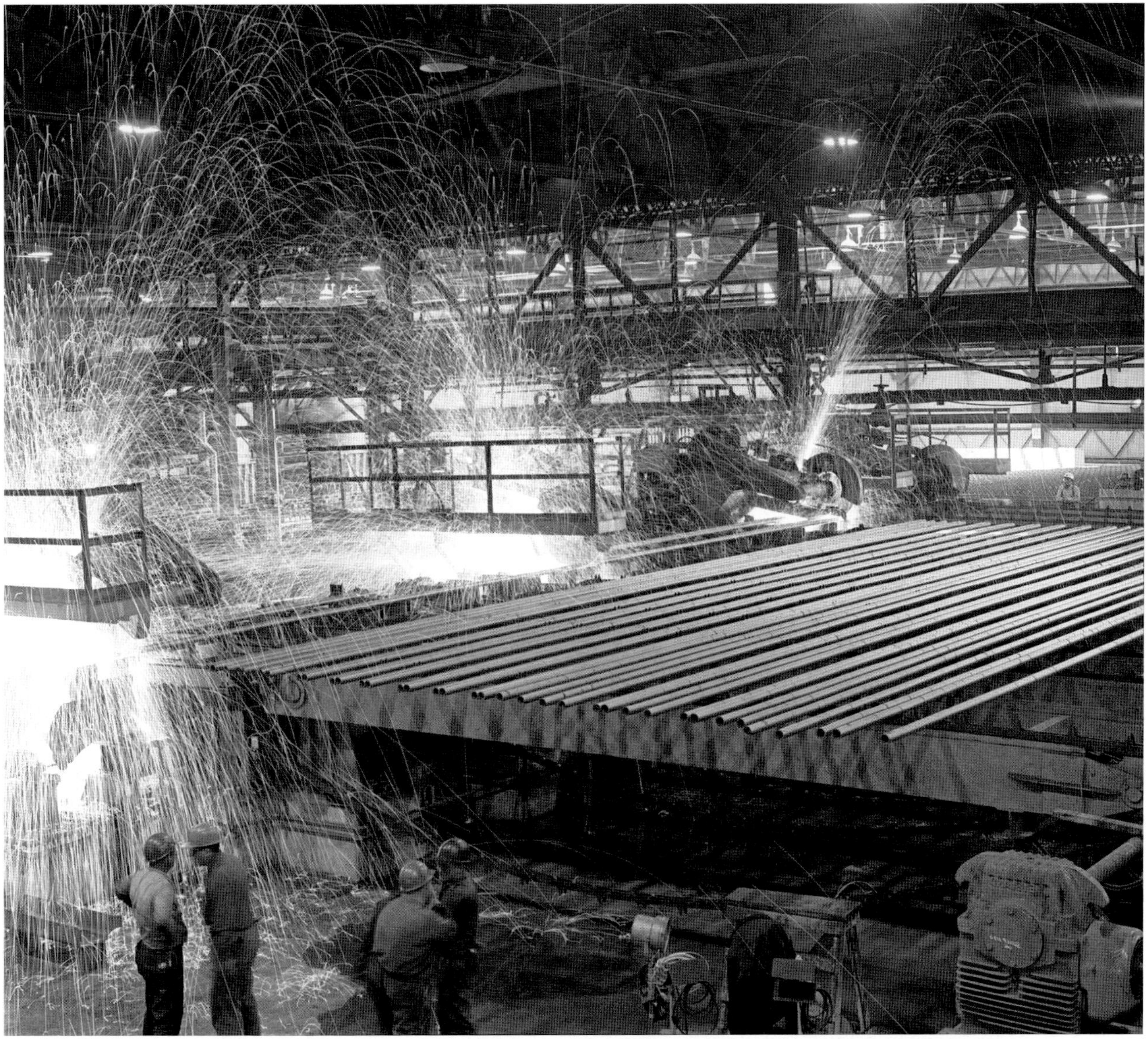

This Youngstown mill specialized in electric-weld pipe. The plant produced a strip of steel that was heated in a furnace, then worked through a series of rolls that bent the sheet into a tube and pushed the edges together. Then the tube passed under an electrode wheel, which melted the edges of the sheet and welded them together. In this 1966 photo, twin saws cut steel pipes to length. (photo 6091D)

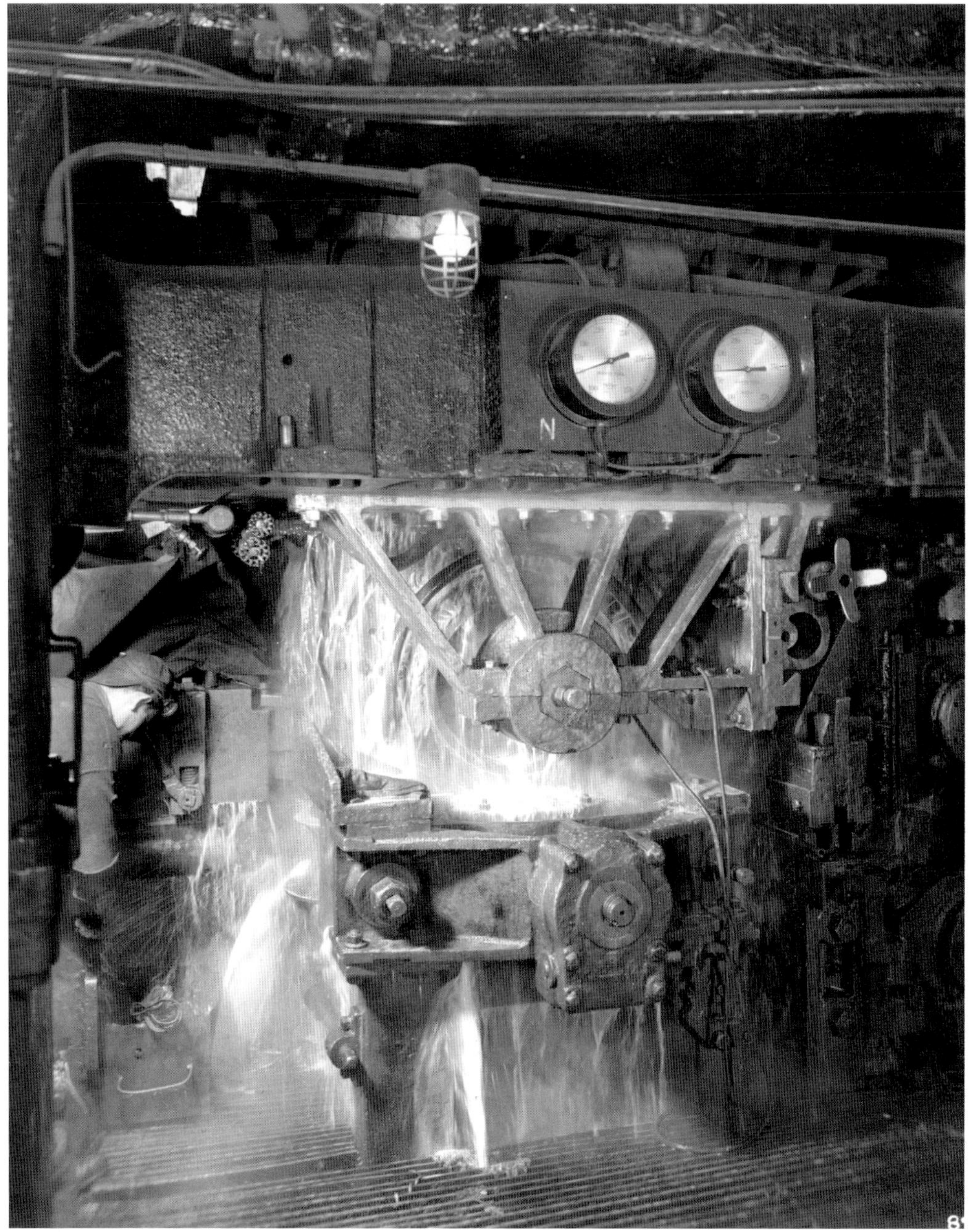

This photo from the Youngstown mill in the 1930s shows the rotary electrodes of the electric-weld pipe mill, which weld the steel sheet into a pipe.

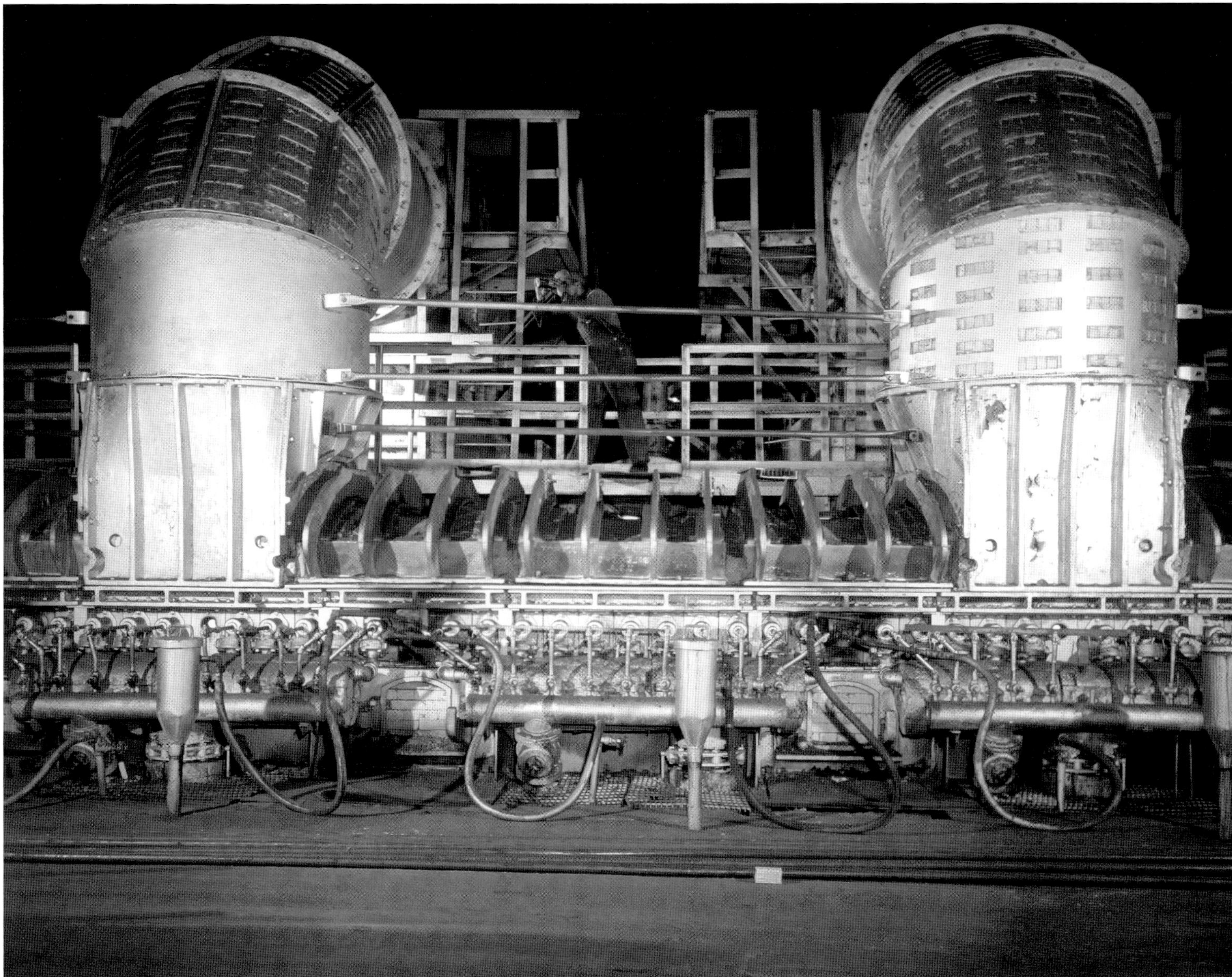

This photo from Youngtown in the 1930s shows the furnace that heated the steel sheets prior to running them through the electric-weld pipe mill. Note the worker in the center of the picture, almost blending into the machinery.

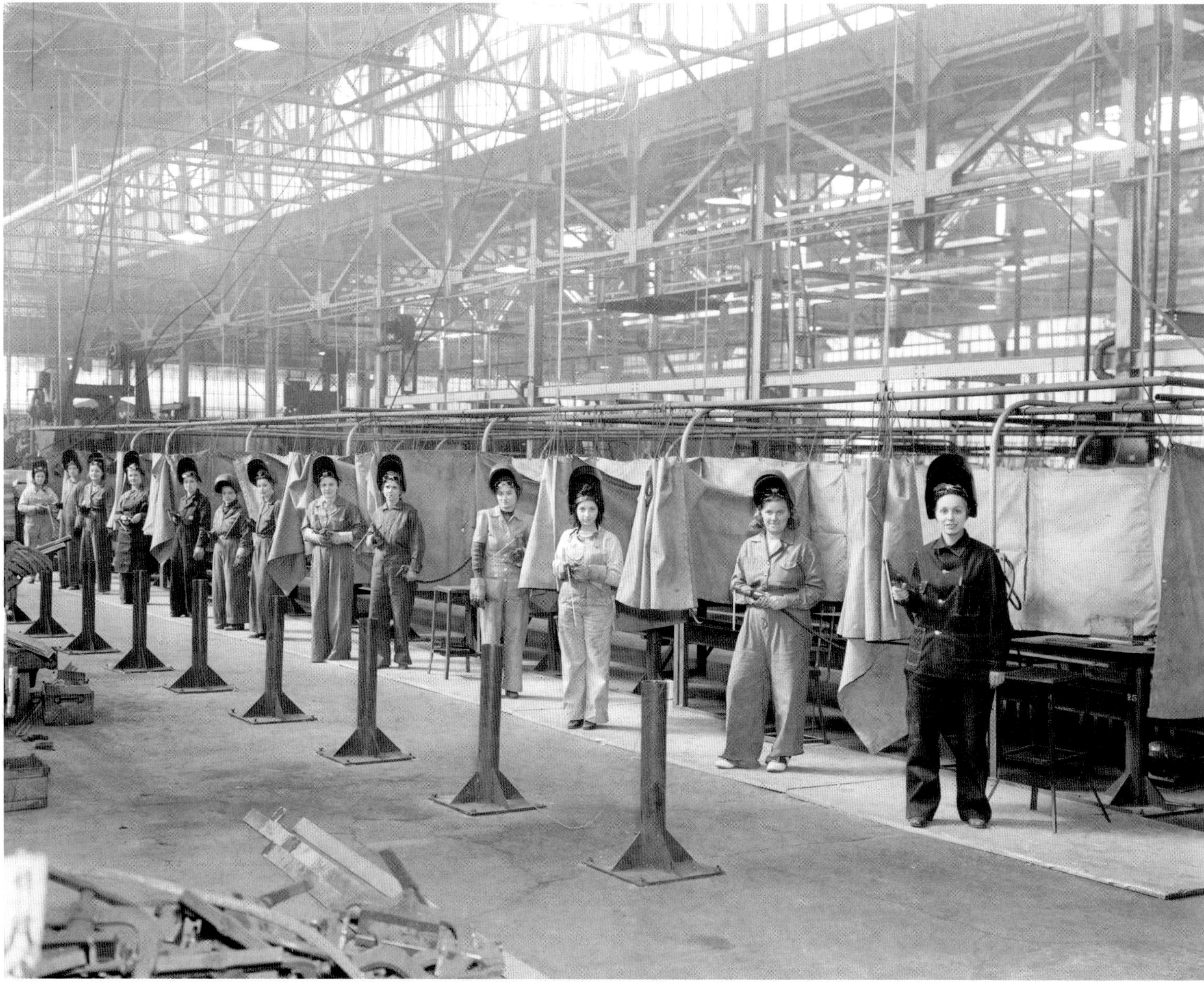

This photo of women workers at Welding School No. 1 was featured in a story about the war effort in the October 1942 issue of *Republic Reports* magazine.

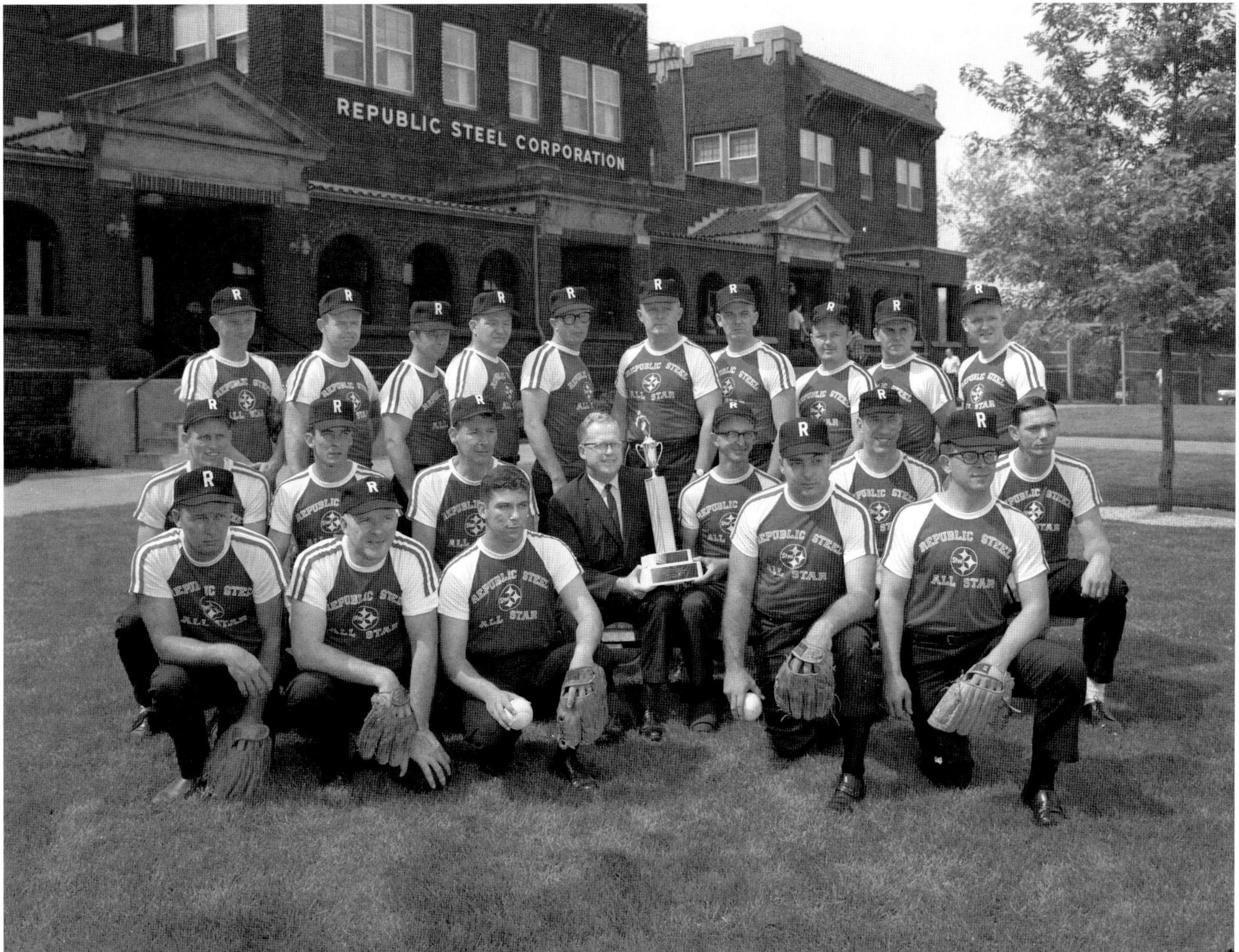

Like many industrial companies, Republic sponsored sports teams. The Warren All-Star softball team poses in front of the main office building, showing off the trophy they won in the 1968 Trumbull County Slow Pitch Tournament, Industrial Division. (photo C643)

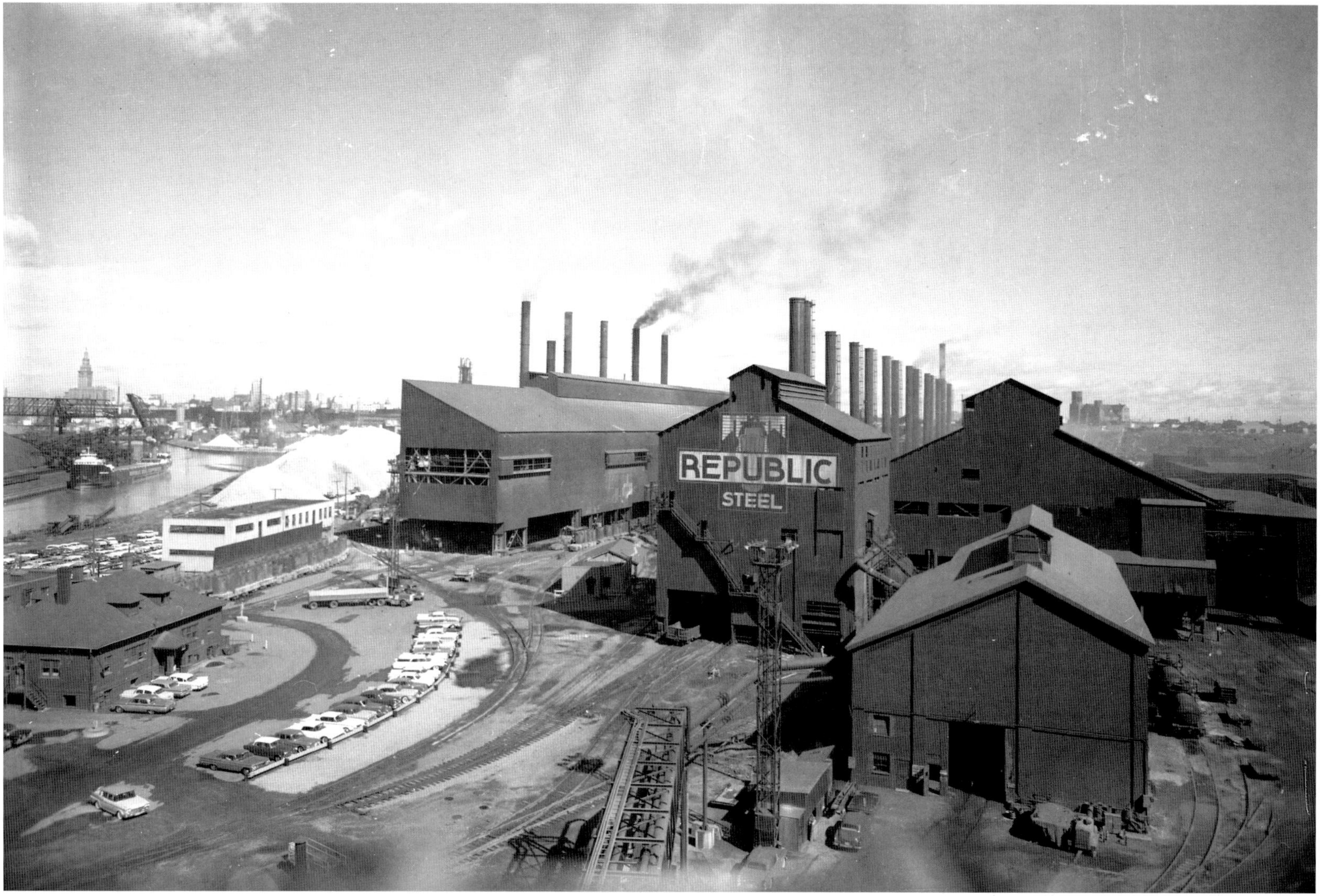

The open hearth furnaces of Republic's Cleveland mill in the 1960s. Built between 1913 and 1916, the former Corrigan, McKinney Steel Company is still in operation today as part of Mittal's Cleveland operation. (Cleveland Engineering Department, Negative 13)

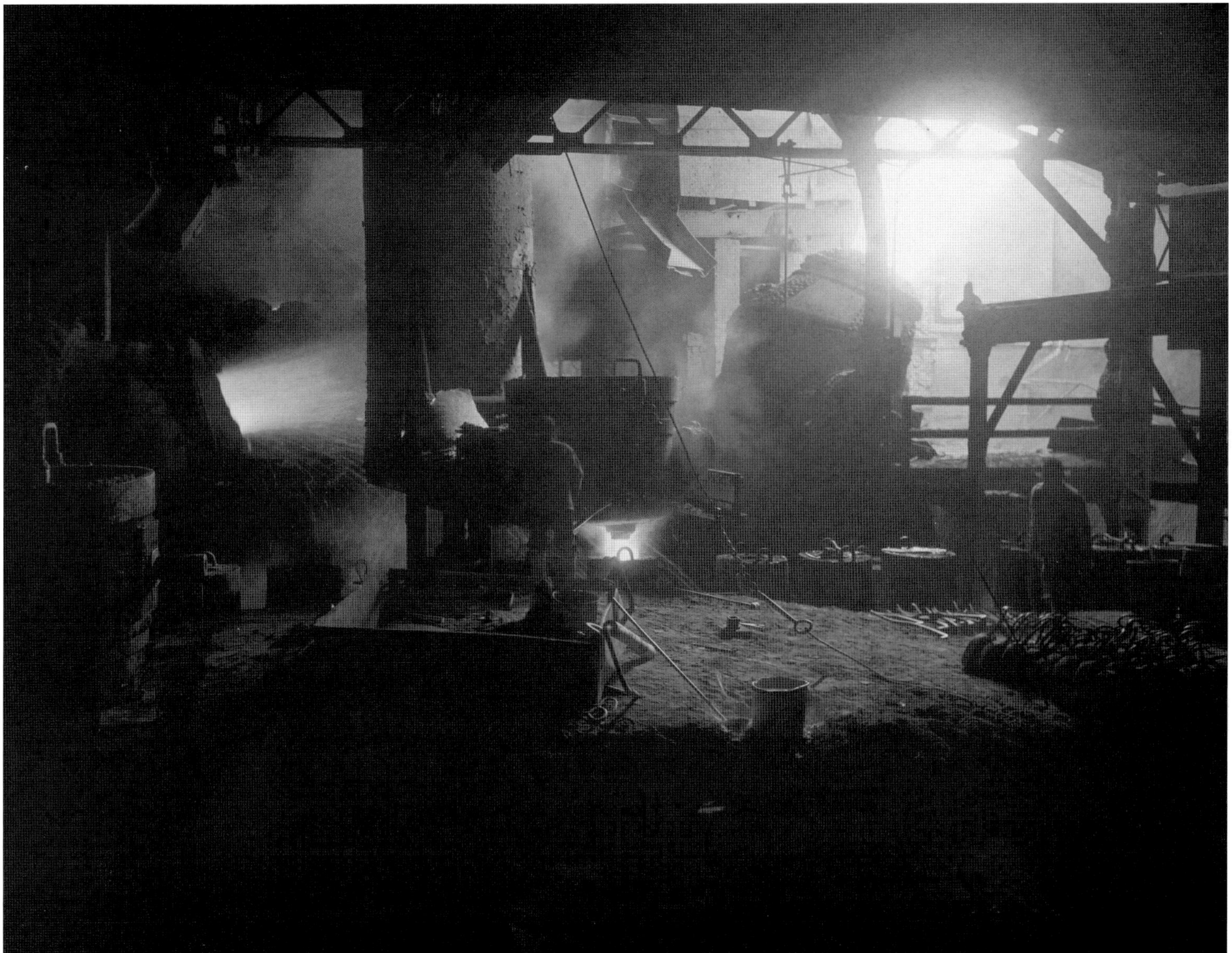

By the time this photo was taken in the 1940s, most steel mills had replaced their Bessemer converters with open hearth furnaces. However, Republic's Youngstown mill, pictured here, kept two Bessemers into the late 1950s for small specialty batches of steel—in this case, for pipes. (Republic Negative C23)

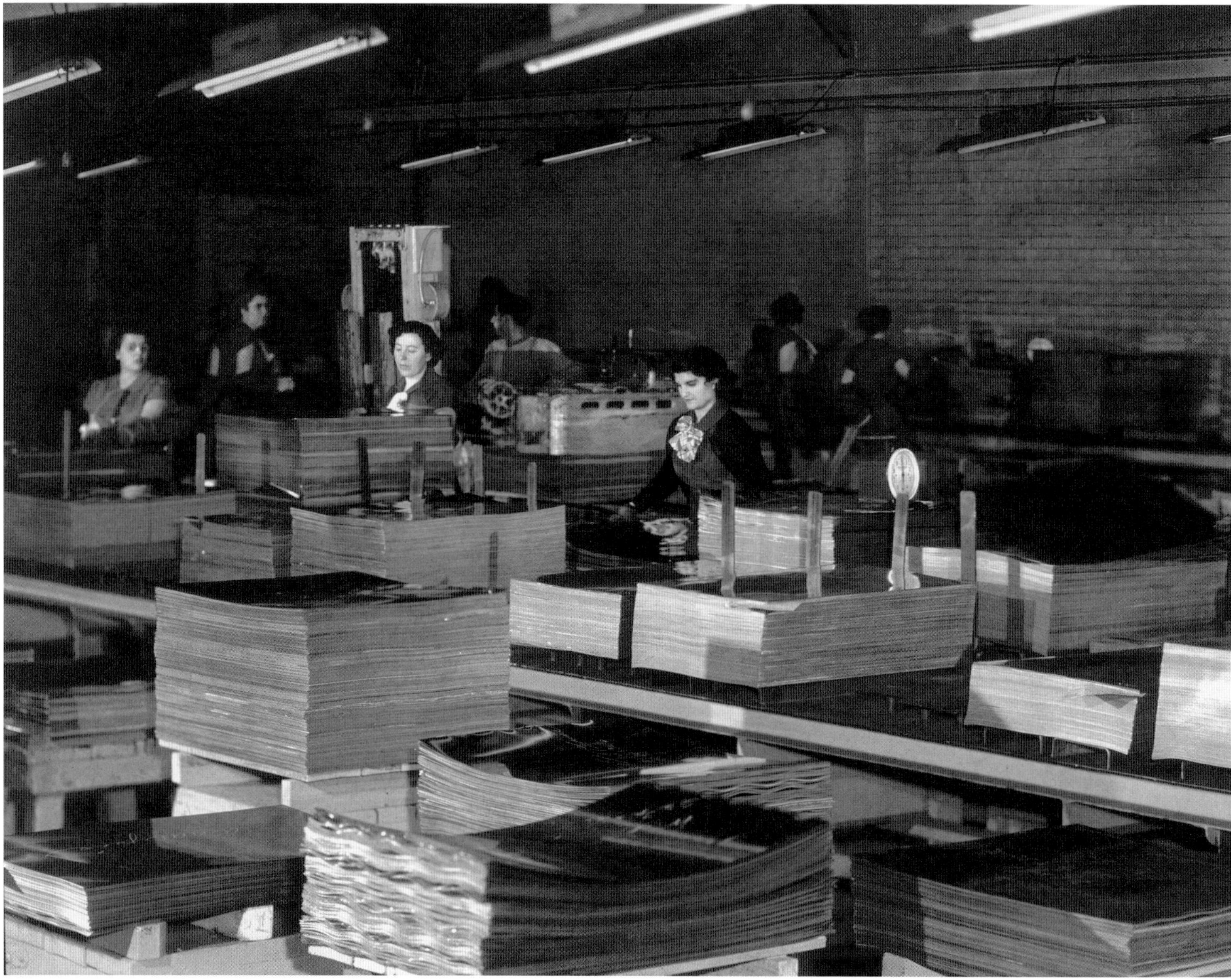

The women in this 1940s photo from the Niles tin mill inspect tin plate, a line of work thought suitable for female employees. The women would glance down the sheet, lift it, and flip it onto another pile to inspect the other side. In many tin mills these women employees became known as "tin flippers" or "tin floppers." (Republic Negative C24)

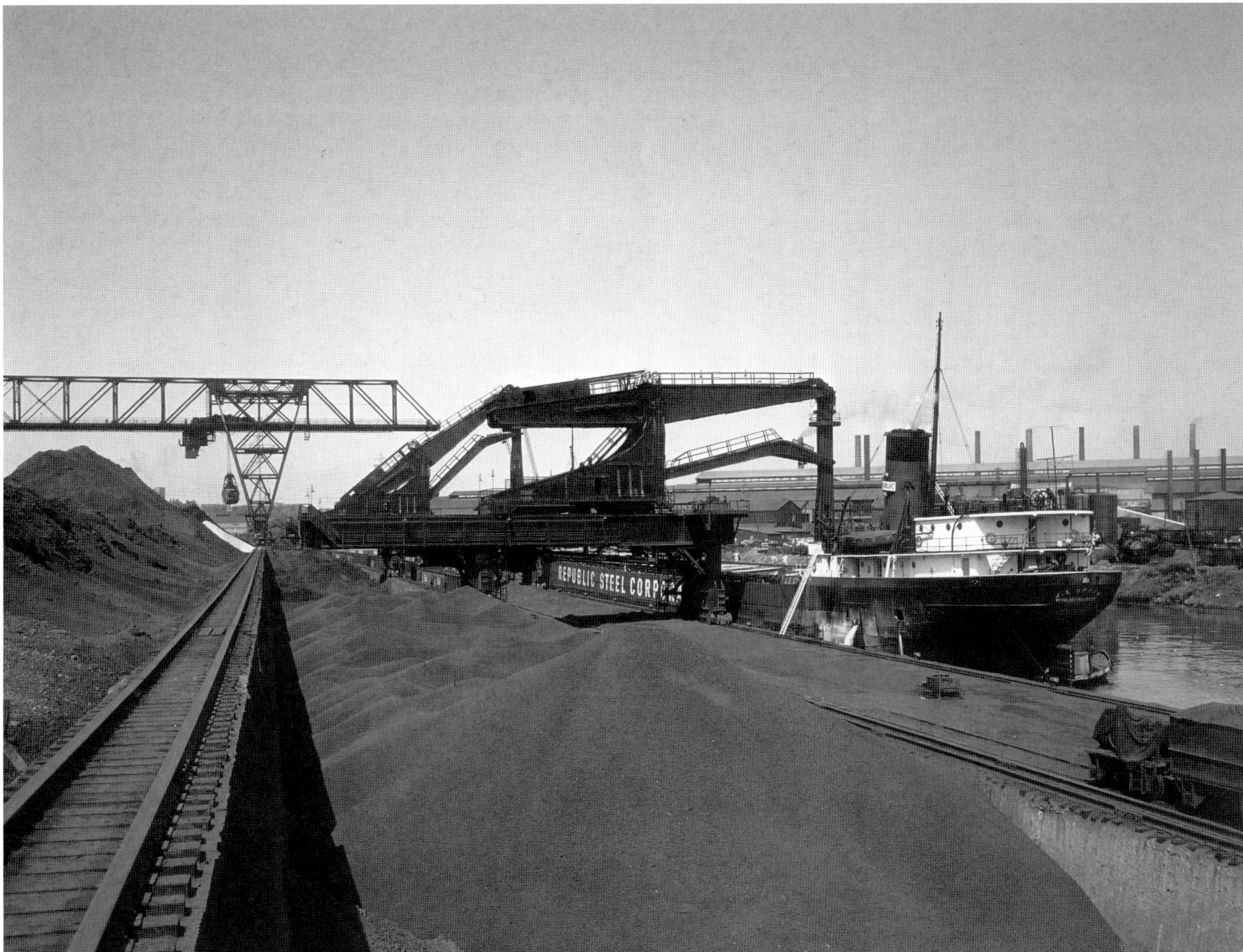

The steamer *J. E. Upson* unloads iron ore at the Republic docks in Cleveland in 1960. Unloading the ore are Hulett unloaders, which reach down into the hold of the ship and remove tons of ore in a single bite. (Republic Negative C180)

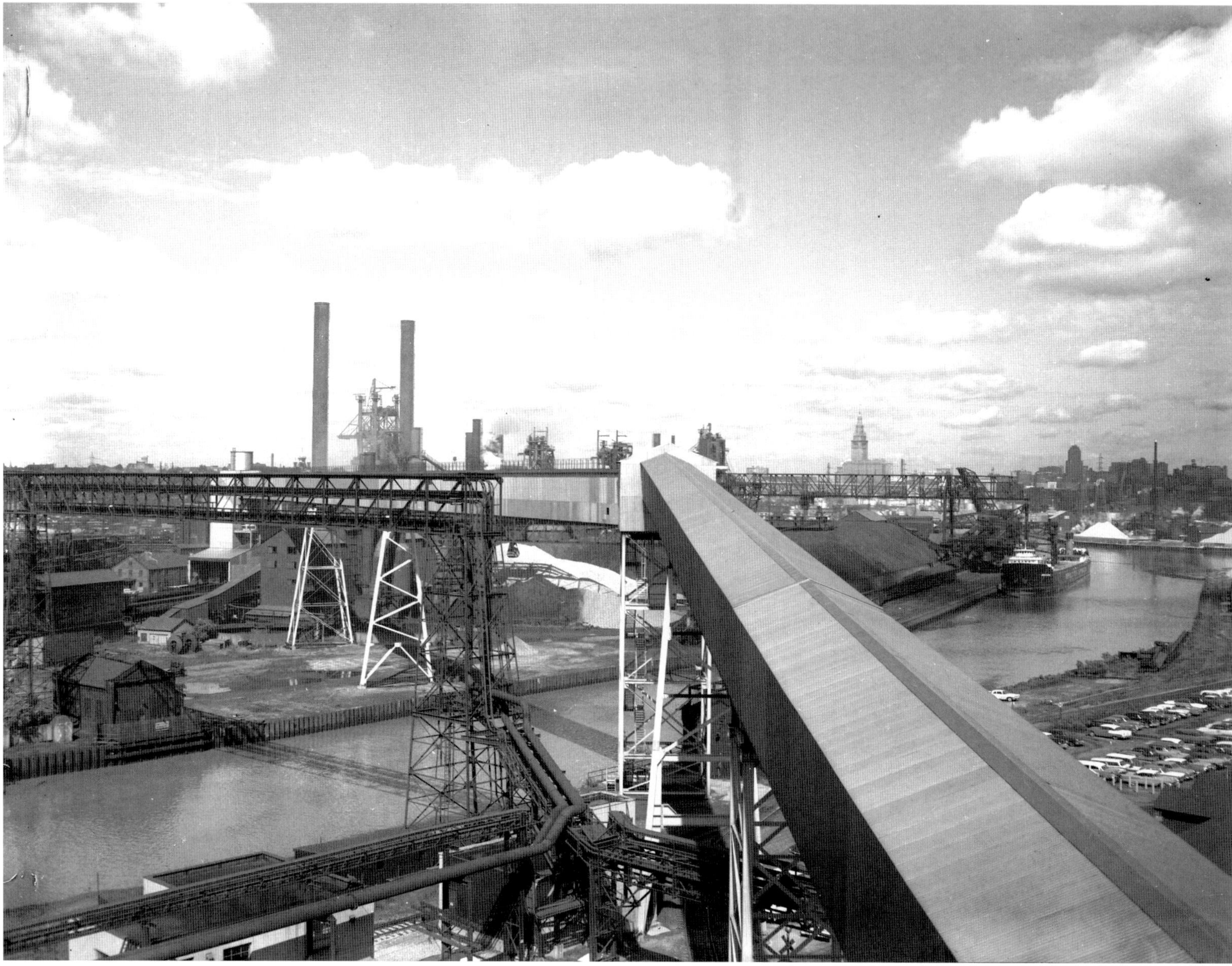

A view of the Cleveland skyline and Republic's blast furnaces from atop the coke conveyor, ca. 1961. (Cleveland District Engineering Negative 5)

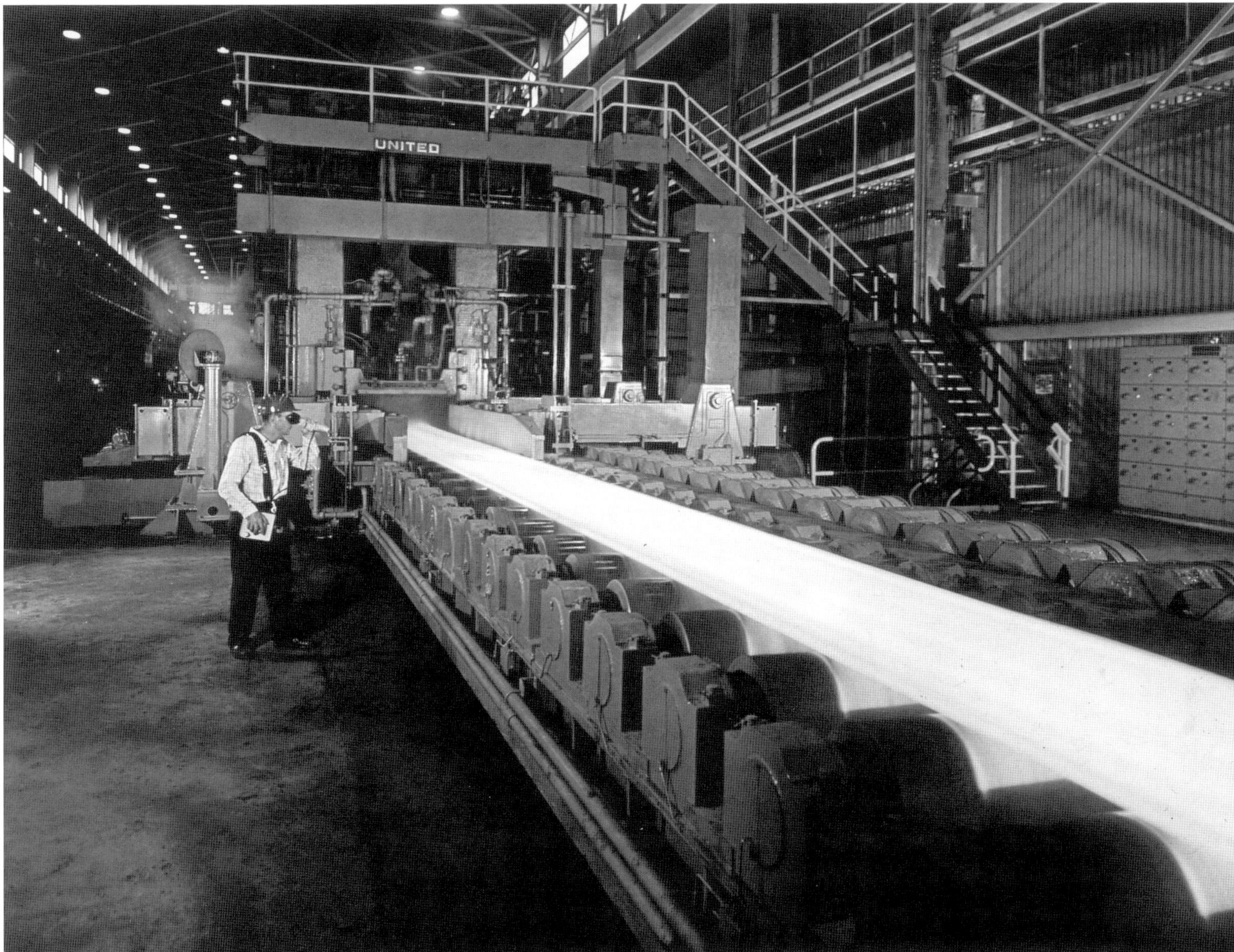

This 1963 photo shows a worker using an electronic device to inspect the hot sheet of steel rolling past in the Warren 56-inch hot strip mill. (Republic Negative C256)

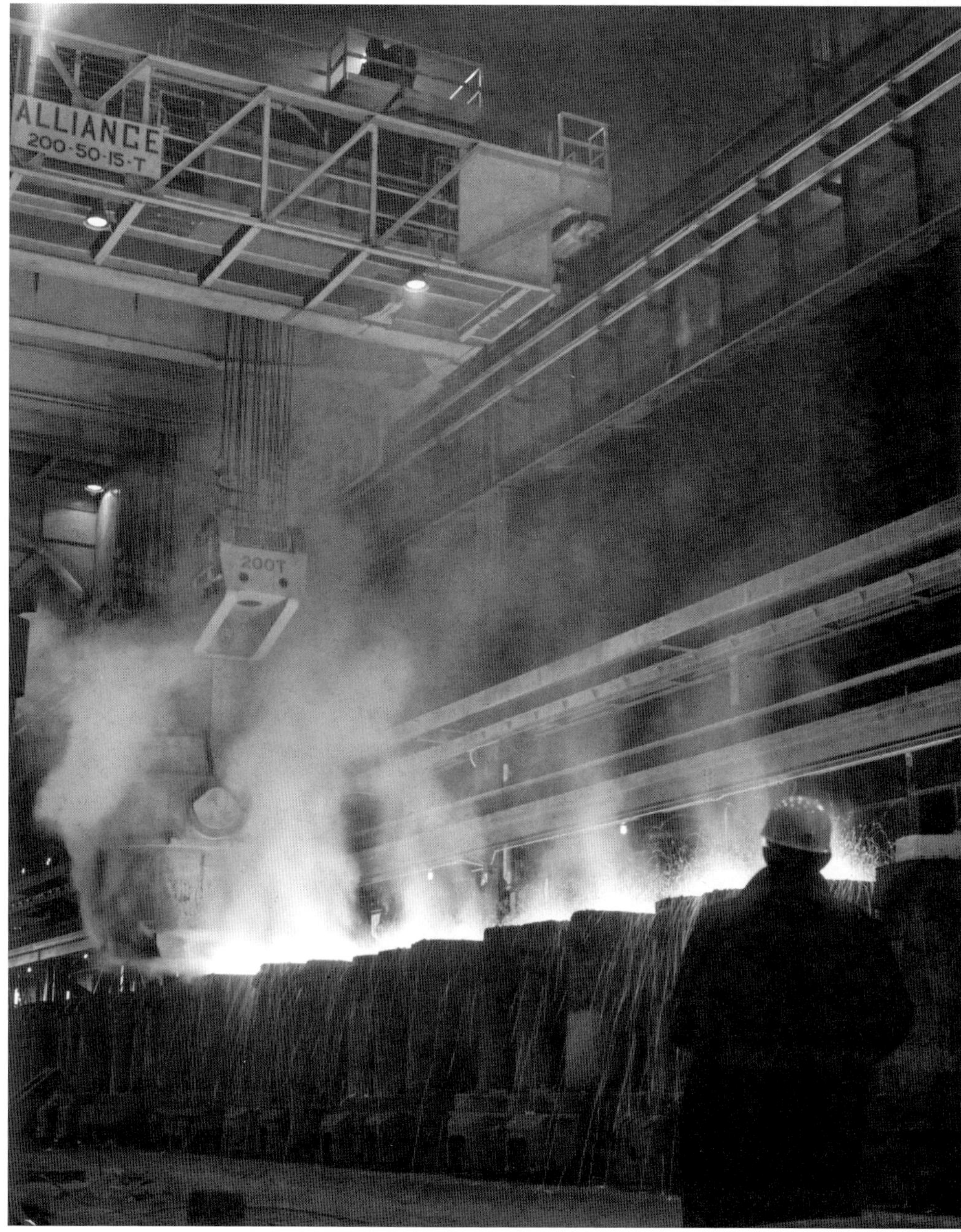

The molten metal, fresh from the Warren plant's new basic oxygen furnace, teems into the ingot molds in this 1966 photo. (Republic Negative C449K)

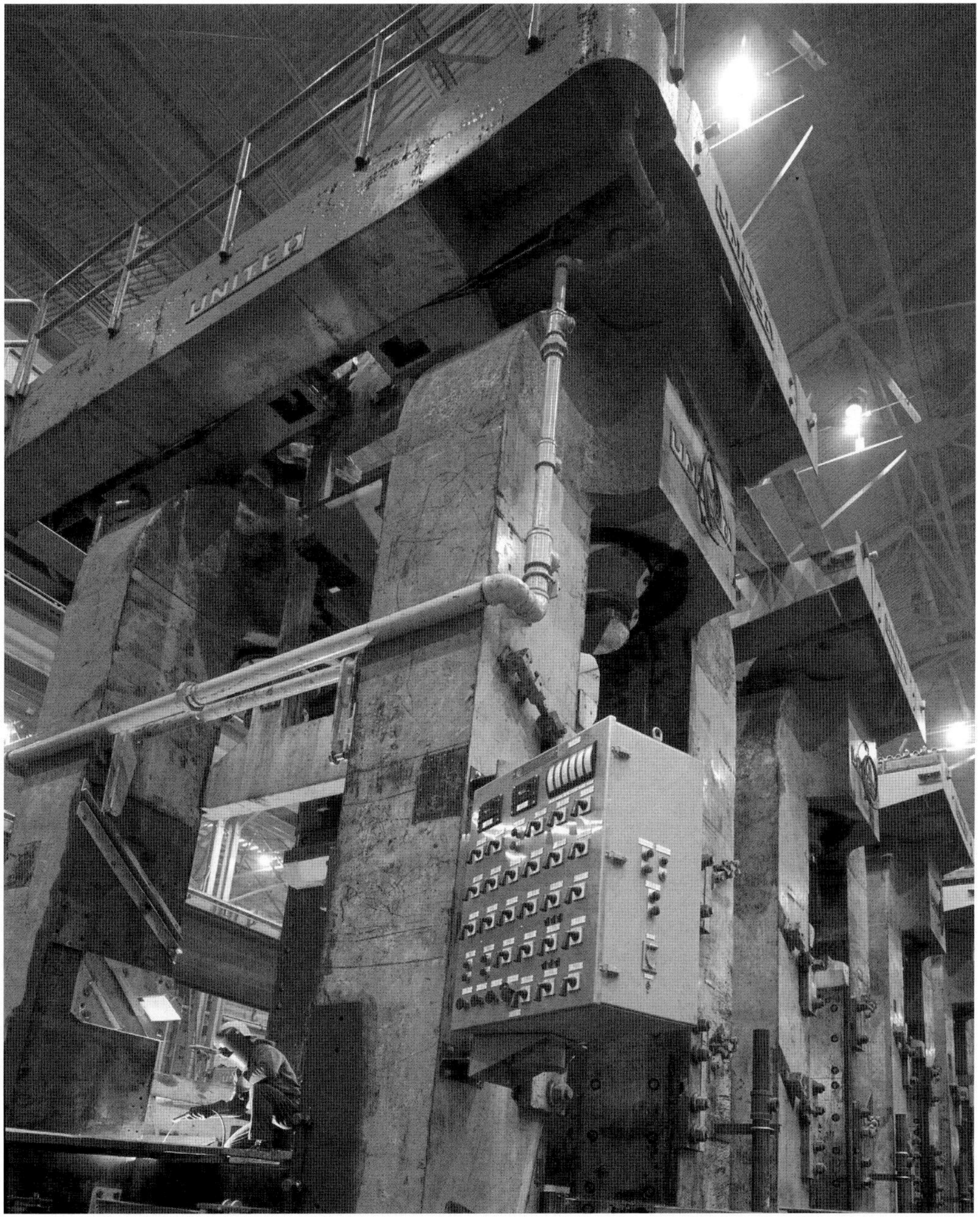

A welder is dwarfed by the gigantic 84-inch cold mill roll stands being installed as part of the Cleveland plant's upgrade in 1969. (Republic Negative C681)

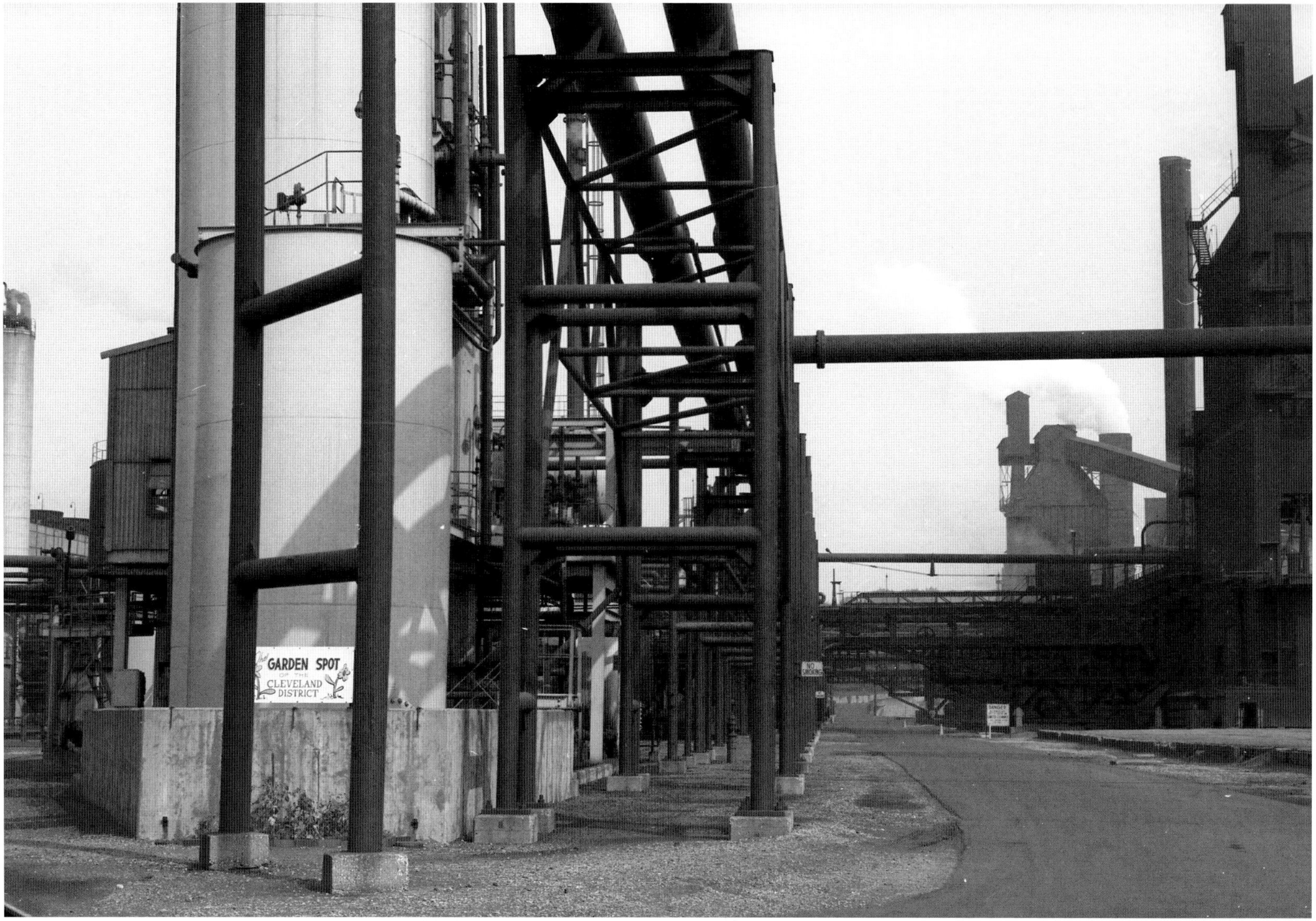

Steelworkers definitely have a sense of humor, as few others would envision a garden spot among the dirt and grime of the coke works. Not only did the workers plant the small garden, but they also had the plant's sign maker construct an appropriate sign in this late 1960s photo. (Cleveland Engineering Department negative, no number)

STEEL PROCESS GLOSSARY

ACID PROCESS: Any steelmaking process in which the furnace lining and the slags are chemically acid.

ADDITIVES: Alloy elements added to the charge, furnace, or ladle to produce steel with required specifications.

ALLOY: Any combination of two or more metals which mix when molten and do not separate.

ANNEALING: A heating process to remove stresses from previous processing and to improve machinability. The process entails heating steel and then cooling it at a set rate to produce the desired strength and formability.

ARC FURNACE: Any furnace in which heat is generated by an electric arc. These furnaces are often referred to as electric arc furnaces.

BASIC OXYGEN FURNACE (BOF): The current method of producing steel. Molten iron from the blast furnace is combined with steel scrap in the BOF vessel. Pure oxygen is blown into the furnace at supersonic speeds to aid combustion, and to refine the iron and scrap through mixing.

BATTERY: The common name for a group of coke ovens.

BESSEMER CONVERTER: A pear-shaped, refractory-lined, steelmaking furnace used in the Bessemer process.

BESSEMER PROCESS: Named after Sir Henry Bessemer, this process involves making steel from pig iron by burning out carbon and other impurities by means of a blast of air forced through the molten metal.

BILLET: 1) A semifinished steel product, rolled or forged from an ingot or strand cast. 2) Square or round steel shape rolled from an ingot or bloom, but smaller than a bloom in cross-section (billets are usually less than 36 square inches in cross-section). Bars, pipes, wire and wire products are made from billets.

BILLET SHEAR: A mechanical shear for cutting billets.

BLAST: The current of air supplied to a blast furnace.

BLAST FURNACE: A device for melting iron ore, using a blast of super-heated air. These are tall, refractory-lined, stack-like furnaces that dominate the skyline of an integrated mill. The furnaces are heated to temperatures in excess of 3,000 degrees (F) by blasts of hot air. Inside are coke, iron ore, and limestone. The coke burns, emitting gases that reduce the iron ore to metallic iron. The limestone combines with impurities in the ore to form slag, which is drained off and discarded when the furnace is tapped.

BLOOM: 1) Wrought-iron blooms, or billets, are formed from the pasty mass remaining when the puddling process is completed. This semi-finished rolled steel weighs approximately 150 pounds and is larger than 5 inches square. 2) A semifinished steel product that has been rolled or forged from an ingot or a strand cast. It usually has a square cross-section exceeding 36 square inches. Blooms are frequently used in the manufacture of building beams and columns.

BLOW: One complete Bessemer cycle from start to finish.

BLOW/BLOWING IN: The process of putting a blast furnace in operation.

BLOWING ENGINE: An engine, originally steam powered, used to accelerate air. These machines put the blast in a blast furnace.

BOF/BOP: Basic oxygen furnace, also called BOP for basic oxygen process.

BY-PRODUCTS OVEN: Coke ovens in which by-product gases are cycled into a commercial recovery and refining system.

CAMPAIGN: The continuous long-term operation of a blast furnace.

CARBON: A nonmetallic element that unites freely with iron and is vital to the chemistry of iron and steel.

CARBON ARC: The electrical arc between two electrodes.

CAST: The release of molten iron into a mold.

CAST HOUSE: Enclosure attached to a blast furnace where iron is first handled in a molten state.

CHARGE: 1) The quantity of iron, scrap, and flux used for one heat. 2) The act of loading the furnace.

CHARGING BOX: 1) A steel box used to deliver scrap, fluxing stone, and other additives in measured quantities in the open hearth, physically about the size of a large bathtub. 2) When used in a BOF, a charging box may be as large as 40 foot. Ruggedly built, it is equipped with fixtures so it can be loaded on a charging machine for delivery to the BOF, or hoisted and tipped by overhead crane for direct charging.

CHARGING DOOR: Insulated doorway used to charge an open hearth furnace.

CHARGING MACHINE: A machine used to charge the open hearth furnace.

COIL: 1) Rolls of thin sheet steel. 2) Rolls of rod or wire. 3) Also refers to the process of rolling or coiling steel.

COKE: A high-carbon fuel made by baking the volatile impurities out of coal, used primarily as blast furnace fuel.

COKE OVENS: A high-temperature oven used to convert coal into coke by baking out volatile materials. An average oven is 18 inches wide, 14 feet high, and 30 feet long.

COLD-ROLLING: After hot-rolling, annealing, and pickling, steel coils are cold-rolled to reduce them to the proper thickness for sale or additional processing.

CONTINUOUS CASTER: A machine equipped with vertical molds called a "strand." Molten steel is teemed into the molds, cooling as it descends, and emerges horizontally as a single slab. Multi-strand machines, with as many as six sets of molds, are used to cast either billets or rounds. This is a much faster method than traditional steelmaking, and completely avoids ingots and blooming mills.

CONTINUOUS FURNACE: A reheating furnace that heats slabs for the hot-roll process.

CROP: (Also a "crop end") The cut-off end of an ingot, billet, or bloom, removed at the first stage of rolling, to square it off for production. This scrap metal is reused for future steel production.

DESCALE: Any process that removes scale during hot-rolling.

DOLOMITE: A form of limestone, used as a refractory in basic steelmaking.

DRILLING: Opening a furnace tap hole, which is pre-drilled and finished with an oxygen lance.

ELECTRIC ARC FURNACES (EAF): A steelmaking furnace that creates heat with an electric arc. This type of furnace is charged with scrap metal (up to 50 percent capacity), limestone, and iron. Many current EAFs only use scrap iron, remelting it to be recast or rerolled. This is the most common type of modern furnace used in minimills.

ELECTRODES: Carbon/graphite rods that provide the arc in an electric arc furnace.

FLUX: Limestone used to remove impurities; forms slag in blast furnaces.

FOUR-HIGH MILL: A rolling mill with four rolls mounted horizontally in each stand. Two work rolls do the actual rolling and shaping, while the other two back up and support the first pair. Technological advancement over the three-high mill.

FRONT WALL: The wall on an open hearth furnace where the charging door is located.

HEARTH: The portion of a furnace which holds the metal being heated or melted.

HEAT: One steelmaking cycle from charging to tapping the molten metal.

HOT BLAST STOVE: Large refractory-lined silos located adjacent to blast furnaces which supply heated air for the blast furnace; used in threes with one working, one regenerating, and one in reserve.

HOT METAL: Molten pig iron or blast furnace iron.

HOT-ROLLING: The process of reheating slabs, billets or blooms and running them through a series of hot mills, where they are reduced to an intermediate thickness and then coiled.

HULETT: A dockside ore-unloading machine found in the lower Great Lakes area, used to transfer iron ore (later taconite) from the holds of ore boats to rail cars. Named after its inventor, George Hulett.

INGOT: Metal that is cast into a mold, weighing as much as 30 tons. Molten steel is poured (or "teemed") from a ladle into an ingot mold. Once it hardens, the mold is removed from the metal, and the steel is rolled or forged into a bloom, billet, or slab.

INGOT MOLD: Cast iron mold that rests on a stool during the teeming of the metal.

INGOT MOLD BUGGY: Small, heavy-duty, four-wheel flat railcar that carries one to four ingot molds during teeming.

INGOT YARD: Area for stockpiling ingots, occasionally while still on buggies, but more often stacked like cordwood.

INTEGRATED: A term used to describe a steel producer that has ironmaking and steelmaking capabilities as well as the ability to process steel into finished products. An integrated producer typically operates a blast furnace to make iron, and has casting, rolling, and other equipment to make semifinished and finished steel products. In the early days of the steel industry, integrated steel companies also owned their own iron ore mines, and often their own ships and rail lines to transport the ore and finished products. The companies were "integrated" because they controlled all the aspects of steelmaking, and thus could control their costs more closely.

LADLE: A large refractory-lined vessel used to move hot metal. A charging ladle is tilted and poured from a spout.

LANCE: A water-cooled pipe to deliver oxygen into a BOF vessel.

LARRY CAR: 1) A car used to charge coke ovens. 2) A part of the Hulett machine which distributes iron ore into waiting railcars.

LIMESTONE: Used as a flux, which combines with impurities in molten metal to form slag.

MELT: A molten charge of metal in an open hearth or BOF.

MELTER: Generic term for furnace operators.

MILL SCALE: Oxides that form on the steel surface during hot-rolling.

OPEN HEARTH PROCESS: A process of making steel from pig iron in a regenerative/reverberatory type of furnace.

ORE BOAT: A special type of freighter used on the Great Lakes to handle iron or taconite. Originally, they were not equipped with unloading machinery, requiring the use of the Hulett ore-unloading machines. Today's ore boats range up to 1,000 feet long and are self-unloading with booms which swing over the dock.

ORE BRIDGE: A bridge-style gantry crane which spans ore, stone, and coal storage piles near a blast furnace.

ORE DOCKS: A very large, elevated structure equipped with multiple tracks designed to hold iron ore or taconite for gravity loading into an ore boat's holds. The ore is stored in numerous pockets, each of which is equipped with a side discharge chute, which are then lowered into the open hatchways of the ore boat's holds for filling.

OXYGEN LANCE: 1) A hand-held tool to clean (scarf) ingot molds. 2) A similar tool used to insert oxygen in an electric melt shop. 3) The water-cooled, oxygen-fed lance used in the BOF.

PICKLING: The action of removing scale or oxides from metal products to obtain a clean surface before further processing.

PIG IRON: A metallic product from the blast furnace containing more than 90 percent iron. It is used directly in the manufacture of steel. In the early days of the iron and steel industry, pig iron was made when the "cinder notch" of the hearth was opened and the molten iron flowed into a main runner from which a worker, using a long bar, directed it into channels cut into sand and arranged like "pigs suckling at the sow" (according to the largely rural-raised workforce in the early iron industry). After the iron cooled, iron breakers, or iron carriers, used heavy sledgehammers and crowbars to break the "pigs" from the "sow," the runner, and to break the iron left in the runner into manageable lengths. Today, the pig iron flows through channels cut into the casting house floor directly into hot metal bottle cars.

PULPIT: An enclosed control position in a steel mill.

QUENCHING: The rapid cooling of incandescent (hot) coke. Also used in the steelmaking process, when hot steel products are sprayed with water or dipped in water baths to rapidly cool them.

RAIL MILL: A rolling mill devoted to rail production.

REFRACTORY: Heat-resistant ceramic materials suitable for furnace lining.

REVERSING MILL: A rolling mill designed to reverse direction following each pass, until the desired reduction is obtained.

ROD MILL: A rolling mill designed for rolling rods.

ROLL: A machined cylinder that actually forms the metal.

ROLL SCALE: A surface oxide formed during rolling, removed using high-pressure water to prevent defects in sheet metal.

ROLLING MILL: In the nineteenth century, rolling mills converted pig iron into semifinished wrought iron bars or plates, which metalworking shops then used in the production of finished goods. After the adoption of the Bessemer and open hearth steelmaking process, rolling mills (often called hot mills) elongated and shaped steel into a semifinished product.

ROUGHING STAND: A mill used for initial rolling of ingots.

RUN OUT: Flat area with powered or live rollers which receives the product after the final rolling pass; also the area that receives billets, rounds, or slabs from a continuous caster.

SCALE: An oxide that forms on the steel surface during heating.

SCARFING: Mechanical smoothing to remove imperfections and scale from ingots or slabs.

SCRAP: 1) Home scrap, includes items such as crop ends, which originate inside the mill. 2) Purchased scrap, obtained from outside firms.

SKIM: The process of holding back the slag floating on the top of the molten iron during tapping of the blast furnace.

SLAB: A semifinished steel product that is hot-rolled down from an ingot or strand cast. It is wide and rectangular in shape, approximately 6 to 9 inches thick and up to 24 to 60 inches wide. It is used in the manufacture of sheets, strip, plates and other flat-rolled steel products.

SLAG: The hot waste drawn off the top of the molten metal processed in a furnace.

SLAG CARS: Specialized rail car designed to carry a thimble-shaped pot used to carry slag.

SLAG DUMP: An area where slag is dumped and cooled so it can be reclaimed and crushed for other uses.

SLAG NOTCH: An opening where slag is tapped from the blast furnace.

SLAG PITS: Pits used to hold and cool hot slag for reprocessing.

STEEL WORKERS ORGANIZING COMMITTEE (SWOC): Labor organization begun by John L. Lewis of the United Mine Workers to engage in the organizing of steelworkers in the 1930s. Once most of the steel companies were unionized by 1942, SWOC changed into the United Steelworkers of America (USWA).

STOPPER ROD: A steel rod covered by refractory tubes that controls the flow of metal teeming from a bottom-pouring ladle.

STRIPPER: Crane that removes ingots from their molds.

STRIPPER CRANES: Heavy-duty combination types used to strip ingot molds using a push-pull action.

TACONITE: A hard-rock form of low-grade iron ore.

TAPHOLE: The opening where iron flows from the blast furnace.

TEEMING: The flow of metal from the ladle to a string of ingot molds.

TEEMING A HEAT: Pouring liquid steel into ingot molds.

TEEMING LADLE: Pours from the bottom into an ingot mold.

THREE-HIGH ROLLING MACHINE: To produce bar, plate, rails, and other shapes, rollers, helpers, and members of the rolling crew manually passed blooms between two metal cylinders rolling in opposite directions. These "rolls" gripped the softened metal and reduced it to the desired shape. A three-high rolling machine allows the workers to pass metal through from both sides, eliminating the time-consuming walk around the machine after each pass. Eventually replaced with the four-high rolling mill, which passed steel blooms through automatically.

TURN: A work period in the steel industry, equivalent to a shift in other industries. In the steel industry for many years, a turn would be twelve hours, and periodically workers had to work a "long turn," a full twenty-four-hour shift, as they switched over from day turn to night turn.

WORKS CONSULTED

Much of the material utilized for the historical chapters of this work came from sources found inside the LTV Steel Collection papers. Particularly for Republic Steel, a number of unpublished company biographies and histories were utilized. These materials have since been cataloged at the Western Reserve Historical Society in Cleveland as part of the collection's processing.

Corrigan, McKinney Steel

"Build Big Furnace on Cuyahoga River." *Cleveland Press,* August 1, 1913.

Chernow, Ron. *Titan: The Life of John D. Rockefeller, Sr.* New York: Random House, 1998. 394–96.

"Corrigan McKinney." *Time* (February 13, 1928).

"Deal for Steel Plant Is Off." *Cleveland Press,* January 29, 1934.

"General Motors Deal Rumored Near for Corrigan McKinney." *Cleveland Press,* January 27, 1934.

Hill, John W. "From Receivership to Many Millions: Amazing Record of McKinney Steel Revealed." *Iron Trade Review* (May 28, 1925): 1379–1417.

Longenecker, Charles. "Corrigan, McKinney Steel Company." *Blast Furnace and Steel Plant* (January 1936): 75–81.

McCoy, Harold G. "Mammoth Rolling Plant in Operation and Dream of Years Becomes Reality." *Cleveland Press,* November 12, 1926.

The Messenger, April 1925.

Miller, Carol Poh. "Corrigan, McKinney Steel Company (Republic Steel)." Library of Congress, Prints and Photographs Division, Historic American Engineering Record, HAER OH-13.

"Republic Steel." *Fortune* (December 1935): 83.

Otis Steel

Bordner, Robert. "Proposed Merger Recalls History of Otis Steel Co." *Cleveland Press,* February 9, 1942.

DeCrane, Ray. "J&L Here Is Moving Ahead Full Blast." *Cleveland Press,* March 11, 1974.

———. "J&L Pouring It On." *Cleveland Press,* April 16, 1974.

"E. J. Kulas: The man that 'peps up steel' as portrayed by Israel Klein, Science Editor NEA Service." *The Otis Sheet* (August 1928).

Ess, T. J. "Jones & Laughlin . . . Cleveland Works." *Iron and Steel Engineer* (February 1959): 86–104.

Heitmann, John A. "Otis Steel Company." In *The Iron and Steel Industry in the Twentieth Century, Encyclopedia of American Business History and Biography.* Ed. Bruce Seely (New York: Facts on File, 1994), 343–44.

Hill, John. "Life Stories of Cleveland Firms, No. 2: The Otis Steel Company." *The Clevelander* (September 1931): 6, 22.

The Otis Sheet, January 1935.

"Quiet Revolution on the Flats." *The Clevelander* (April 1966).

Rockwell, Guy T. "Otis Seeks Tie to Jones & Laughlin." *Cleveland Plain Dealer,* February 8, 1942.

"Sixty Cities Seek Otis Steel Plant." *The Cleveland Press,* March 17, 1913.

"60,000 Will Be Brought to City by Steel Mills." *The Cleveland Press,* January 1, 1914.

Wagman, Jules. "Steel Town: J&L Leader Tells of Strengths That Will Keep Industry Here." *The Cleveland Press,* December 1, 1979.

Wollman, David H., and Donald R. Inman. *Portraits in Steel: An Illustrated History of Jones & Laughlin Steel Corporation.* Kent, Ohio: Kent State University Press, 1999.

Republic Steel

Cook, Philip L. "Tom M. Girdler and the Labor Policies of Republic Steel Corporation." *Social Science* (January 1967): 21–30.

"Eaton Steel Merger Wins." *Cleveland Plain Dealer,* December 18, 1929.

"Elmer T. McCleary. Dead: Head of Republic Dies Unexpectedly." *Youngstown Telegram,* April 22, 1930.

Fast, Howard. "An Occurance at Republic Steel." In *The Aspirin Age, 1914–1941.* Ed. Isabel Leighton. New York: Simon and Schuster, 1949.

Girdler, T. M. biography, Republic Steel file. LTV Steel Collection.

———. *Bootstraps: The Autobiography of Tom M. Girdler.* New York: Charles Scribner's Sons, 1943.

———. "Why Cleveland Was Chosen." *The Clevelander* (September 1937).

Greenough, Peter B. "Republic Makes Ready to Tap Giant Furnace." *Cleveland Plain Dealer,* November 25, 1949.

H. A. Brassert & Company. "Report on the Proposed Consolidation of Republic Iron and Steel Company, Central Alloy Steel Company, Donner Steel Company, Bourne-Fuller Company for Cyrus Eaton." Chicago: H. A. Brassert & Company, 1929.

Hoag, David H. *The LTV Corporation: The Evolution of a Company and an Industry.* New York: The Newcomen Society of the United States, 1995.

Hurd, H. M. "Republic History." Youngstown, Ohio: Republic Iron and Steel Company, 1918.

"Introduction to Republic Steel and the Steel Industry." Cleveland, Ohio: Republic Steel Corporation, 1953.

"Major Expansion Stretches Steel Mill to Limit." *Engineering News-Record* (September 20, 1951): 30–33.

Miller, Carol Poh. "Republic Steel." In *The Iron and Steel Industry in the Twentieth Century, Encyclopedia of American Business History and Biography.* Ed. Bruce Seely (New York: Facts on File, 1994), 366–69.

"The New LTV Steel." *33 Metal Producing* (August 1993).

Patton, Thomas F. "Recollections of Some of the Highlights of the History of Republic Steel Corporaton, 1930–1971." Cleveland, Ohio: Republic Steel Corporation, 1981.

"Republic." *Fortune* (December 1935): 76–84, 142, 144, 147–48, 150, 152.

"Republic Steel: Building toward the 80s." *33 Metal Producing* (January 1979).

Republic Steel Corporation. *Republic Goes to War.* Cleveland, Ohio: Republic Steel Corporation, 1945.

"Republic Steel Fact Book." Cleveland, Ohio: Republic Steel Corporation, 1959.

"The Republic Steel Story." *33 Magazine* (August 1971).

Rose, Marc. "Seven Years of Republic Steel." Cleveland, Ohio: Republic Steel Corporation, 1937.

Smith, Earle C. "History of Republic Steel Corporation." Cleveland, Ohio: Republic Steel Corporation, 1936.

"Steelmaking, Central Alloy Division." Massillon, Ohio: Republic Steel Corporation, n.d.

"Steelmaking at Republic." Cleveland, Ohio: Republic Steel Corporation, 1948.

"Supporting Data for Army-Navy Production Award." Massillon, Ohio: Republic Steel Corporation, 1942.

"A Talk with Cyrus Eaton." *Forbes* (March 1, 1968).

"350,000,000 Deal Links 5 Steel Groups." *The Cleveland News,* December 18, 1929.

Wellman, Bertha. "Republic Steel Readies Giant New Open Hearth." *Cleveland Press,* August 30, 1949.

INDEX